Progress in Nonlinear Differential Equations and Their Applications
Volume 13

Fabrice Bethuel
Haïm Brezis
Frédéric Hélein

Ginzburg-Landau Vortices

Springer Science+Business Media, LLC

Fabrice Bethuel
Laboratoire d'Analyse Numérique
Université Paris-Sud
91405 Orsay Cedex
France

Haïm Brezis
Analyse Numérique
Université Pierre et Marie Curie
4, place Jussieu
75252 Paris Cedex 05, France
and
Department of Mathematics
Rutgers University
New Brunswick, NJ 08903

Frédéric Hélein
CMLA, ENS-Cachan
94235 Cachan Cedex
France

Library of Congress Cataloging-in-Publication Data

Bethuel, Fabrice, 1963-
Ginzburg-Landau vortices / Fabrice Bethuel, Haïm Brezis, Fréderic Hélein.
p. cm. -- (Progress in nonlinear differential equations and their applications ; v. 13)
Included bibliographical references and index.
ISBN 978-0-8176-3723-1 ISBN 978-1-4612-0287-5 (eBook)
DOI 10.1007/978-1-4612-0287-5
1. Singularities (Mathematics) 2. Mathematical physics.
3. Superconductors--Mathematics. 4. Superfluidity--Mathematics.
5. Differential equations, Nonlinear--Numberical solutions.
I. Brezis, H. (Haim) II. Hélein, Fréderic, 1963- . III. Title.
IV. Series.
QC20.7.S54B48 1994 94-2026
530.1'55353--dc20 CIP

Printed on acid-free paper

ISBN 978-0-8176-3723-1

Typeset by the Authors in AMSTEX.

9 8 7 6 5 4 3 2

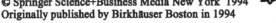

TABLE OF CONTENTS

Acknowledgements

We are grateful to E. DeGiorgi, H. Matano, L. Nirenberg and L. Peletier for very stimulating discussions. During the preparation of this work we have received advice and encouragement from many people: A. Belavin, E. Brezin, N. Carlson, S. Chanillo, B. Coleman, L.C. Evans, J. M. Ghidaglia, R. Hardt, B. Helffer, M. Hervé, R. M. Hervé, D. Huse, R. Kohn, J. Lebowitz, Y. Li, F. Merle, J. Ockendon, Y. Pomeau, T. Rivière, J. Rubinstein, I. Shafrir, Y. Simon, J. Taylor and F. Treves.

Part of this work was done while the first author (F.B.) and the third author (F.H.) were visiting Rutgers University. They thank the Mathematics Department for its support and hospitality; their work was also partially supported by a Grant of the French Ministry of Research and Technology (MRT Grant 90S0315). Part of this work was done while the second author (H.B.) was visiting the Scuola Normale Superiore of Pisa; he is grateful to the Scuola for its invitation. We also thank Lisa Magretto and Barbara Miller for their enthusiastic and competent typing of the manuscript.

INTRODUCTION

The original motivation of this study comes from the following questions that were mentioned to one of us by H. Matano.

Let

$$G = B_1 = \{x = (x_1, x_2) \in \mathbf{R}^2;\ x_1^2 + x_2^2 = |x|^2 < 1\}.$$

Consider the Ginzburg-Landau functional

$$E_\varepsilon(u) = \frac{1}{2}\int_G |\nabla u|^2 + \frac{1}{4\varepsilon^2}\int_G (|u|^2 - 1)^2 \tag{1}$$

which is defined for maps $u \in H^1(G;\mathbb{C})$ also identified with $H^1(G;\mathbf{R}^2)$.

Fix the boundary condition

$$g(x) = x \quad \text{on} \quad \partial G$$

and set

$$H_g^1 = \{u \in H^1(G;\mathbb{C});\quad u = g \quad \text{on} \quad \partial G\}.$$

It is easy to see that

$$\operatorname*{Min}_{u \in H_g^1} E_\varepsilon(u) \tag{2}$$

is achieved by some u_ε that is smooth and satisfies the Euler equation

$$\begin{cases} -\Delta u_\varepsilon = \dfrac{1}{\varepsilon^2} u_\varepsilon(1 - |u_\varepsilon|^2) & \text{in } G, \\ u_\varepsilon = g & \text{on } \partial G. \end{cases} \tag{3}$$

The maximum principle easily implies (see e.g., F. Bethuel, H. Brezis and F. Hélein [2]) that any solution u_ε of (3) satisfies $|u_\varepsilon| \le 1$ in G. In particular, a subsequence (u_{ε_n}) converges in the $w^\star - L^\infty(G)$ topology to a limit $u^\star$. Clearly, $|u^\star(x)| \le 1$ a.e. It is very easy to prove (see Chapter III) that

$$\int_G (|u_\varepsilon|^2 - 1)^2 \le C\varepsilon^2 |\log \varepsilon| \tag{4}$$

and thus $|u_{\varepsilon_n}(x)| \to 1$ a.e. This suggests that $|u^\star(x)| = 1$ a.e. However, such a claim is not clear at all since we do not know, at this stage, that $u_{\varepsilon_n} \to u^\star$ a.e. It turns out to be true that $|u^\star(x)| = 1$ a.e. — but we have no simple proof. This fact is derived as a consequence of a delicate analysis (see Chapter VI).

The original questions of H. Matano were:

Question 1: Does $\lim_{\varepsilon \to 0} u_\varepsilon(x)$ exist a.e.?

Question 2: What is u^*? Do we have $u^*(x) = x/|x|$?

Question 3: What can be said about the zeroes of u_ε? If they are isolated do they have degree ± 1 (in the sense of Section IX.1)?

These questions have prompted us to consider a more general setting. Let $G \subset \mathbb{R}^2$ be a smooth, bounded and simply connected domain in $\mathbb{R}^2$. Fix a (smooth) boundary condition $g : \partial G \to S^1$ and consider a minimizer u_ε of problem (2) as above. Our purpose is to study the behavior of u_ε as $\varepsilon \to 0$.

The Brouwer degree

$$d = \deg(g, \partial G) \tag{5}$$

(i.e., the winding number of g considered as a map from ∂G into S^1) plays a crucial role in the asymptotic analysis of u_ε.

Case d = 0. This case is easy because $H^1_g(G; S^1) \neq \emptyset$ and thus the minimization problem

$$\operatorname*{Min}_{u \in H^1_g(G;S^1)} \int_G |\nabla u|^2 \tag{6}$$

makes sense. In fact, problem (6) has a unique solution u_0 that is a smooth harmonic map from G into S^1, i.e.,

$$-\Delta u_0 = u_0 |\nabla u_0|^2 \quad \text{in} \quad G.$$

Moreover (see e.g., Lemma 1 in F. Bethuel, H. Brezis and F. Hélein [2])

$$u_0 = e^{i\varphi_0} \quad \text{in} \quad G$$

where φ_0 is a harmonic function (unique mod $2\pi\mathbb{Z}$) such that

$$e^{i\varphi_0} = g \quad \text{on} \quad \partial G.$$

We have proved in F. Bethuel, H. Brezis and F. Hélein [2] (see also Appendix I at the end of the book) that $u_\varepsilon \to u_0$ in $C^{1,\alpha}(\overline{G})$ and in $C^k_{\text{loc}}(G)$ $\forall k$; in particular,

$$\int_G |\nabla u_\varepsilon|^2 \text{ remains bounded as } \varepsilon \to 0. \tag{7}$$

We have also obtained rates of convergence for $\|u_\varepsilon - u_0\|$ in various norms.

Case d≠0. Throughout the book we assume that $d > 0$ since the case $d < 0$ reduces to the previous case by complex conjugation. Here, the main difficulty stems from the fact that

$$H_g^1(G; S^1) = \emptyset. \tag{8}$$

Indeed, suppose not, and say that $H_g^1(G; S^1) \neq \emptyset$, then we could consider, as above,

$$\operatorname*{Min}_{H_g^1(G;S^1)} \int_G |\nabla u|^2. \tag{9}$$

A minimizer exists and is smooth up to ∂G, e.g., by a result of C. Morrey [1],[2]. In particular, there would be some $u \in C(\overline{G}; S^1)$ such that $u = g$ on ∂G. Standard degree theory shows that this is impossible since g can be homotopied in S^1 to a constant. Alternatively, one could also use $H^{1/2}(S^1; S^1)$ degree theory (see a result of L. Boutet de Monvel and O. Gabber quoted in A. Boutet de Monvel-Berthier, V. Georgescu and R. Purice [1], [2]) to show that $H_g^1(G; S^1) = \emptyset$.

In this case, problem (9) does not make sense. In order to get around this topological obstruction we are led to the following idea. Enlarge the class of testing functions to

$$H_g^1(G; \mathbb{C}).$$

(Clearly this set is always nonempty.) But on the other hand, add a penalization in the energy that "forces" $|u|$ to be close to 1. The simplest penalty that comes to mind is

$$\frac{1}{\varepsilon^2} \int_G (|u|^2 - 1)^2.$$

Therefore, we are led very naturally to

$$\operatorname*{Min}_{H_g^1(G;\mathbb{C})} E_\varepsilon.$$

Here, in contrast with the previous case,

$$\int_G |\nabla u_\varepsilon|^2 \to +\infty, \quad \text{as } \varepsilon \to 0, \tag{10}$$

(otherwise, $u_{\varepsilon_n} \rightharpoonup \tilde{u}$ weakly in H^1 and $u_{\varepsilon_n} \to \tilde{u}$ a.e., so that $|\tilde{u}| = 1$, a.e.; thus $\tilde{u} \in H_g^1(G; S^1)$ — impossible by (8)). However, we may still hope that

$$u_\star(x) = \lim\ u_{\varepsilon_n}(x) \text{ exists for a.e. } x \in G$$

(naturally, with $\int_G |\nabla u_\star|^2 = \infty$). If this is indeed the case then $u_\star$ can be viewed as a "generalized solution" of problem (9).

Of course, many other "penalties" can be devised. They all seem to lead to the same class of generalized solutions. For example, one other natural penalty consists of drilling a few little holes $B(a_i, \rho)$ in G and considering the domain $G_\rho = G \setminus \bigcup_i B(a_i, \rho)$. In this case there is no topological obstruction and

$$H_g^1(G_\rho; S^1) \neq \emptyset$$

(we do not impose a Dirichlet condition on $\partial B(a_i, \rho)$). Then, one may consider the problem

$$\underset{H_g^1(G_\rho;S^1)}{\text{Min}} \int_{G_\rho} |\nabla u|^2$$

and analyze what happens as $\rho \to 0$. Here, the points (a_i) are free to move and some configurations will turn out to be "better" than others (see Section I.4 and Chapter VIII).

Going back to a minimizer u_ε of the original functional E_ε, our main results are the following:

Theorem 0.1. *Assume G is starshaped. Then there is a subsequence $\varepsilon_n \to 0$ and exactly d points $a_1, a_2, \ldots, a_d$ in G and a smooth harmonic map $u_\star$ from $G \setminus \{a_1, a_2, \ldots, a_d\}$ into S^1 with $u_\star = g$ on ∂G such that*

$$u_{\varepsilon_n} \to u_\star \text{ in } C^k_{\text{loc}}(G \setminus \bigcup_i \{a_i\}) \ \forall k \text{ and in } C^{1,\alpha}(\bar{G} \setminus \bigcup_i \{a_i\}) \ \forall \alpha < 1.$$

In addition, each singularity has degree +1 and, more precisely, there are complex constants (α_i) with $|\alpha_i| = 1$ such that

$$\left| u_\star(z) - \alpha_i \frac{(z - a_i)}{|z - a_i|} \right| \leq C|z - a_i|^2 \text{ as } z \to a_i, \ \forall i. \tag{11}$$

This theorem answers, in particular, Question 1 above. In this theorem it is essential (in general) to pass to a subsequence. For example, if G is the unit disc and $g = e^{2i\theta}$ then, for ε small, u_ε is **not** unique and various subsequences converge to different limits (see Section VIII.5). However, in some cases, for example $g(\theta) = e^{i\theta}$, the full sequence (u_ε) converges to a well defined limit (see Section VIII.4).

So far, we have not said anything about the location of the singularities. Our next result tells us where to find them. For this purpose, we introduce, for any given configuration $b = (b_1, b_2, \ldots, b_d)$ of distinct points in G, the function

$$W(b) = -\pi \sum_{i \neq j} \log |b_i - b_j| + \frac{1}{2} \int_{\partial G} \Phi(g \times g_\tau) - \pi \sum_{i=1}^{d} R(b_i) \tag{12}$$

where Φ is the solution of the linear Neumann problem

$$\begin{cases} \Delta \Phi = 2\pi \sum_{i=1}^{d} \delta_{b_i} & \text{in } G, \\ \dfrac{\partial \Phi}{\partial \nu} = g \times g_\tau & \text{on } \partial G, \end{cases}$$

(ν is the outward normal to ∂G and τ is a unit tangent vector to ∂G such that (ν, τ) is direct) and

$$R(x) = \Phi(x) - \sum_{i=1}^{d} \log |x - b_i|.$$

Note that $R \in C(\overline{G})$, so that $R(b_i)$ makes sense.

The function W, called the **renormalized energy**, has the following properties (see Section I.4):

(i) $W \to +\infty$ as two of the points b_i coalesce,

(ii) $W \to +\infty$ as one of the points b_i tends to ∂G (since $R(b_i) \to -\infty$ as $b_i \to \partial G$).

In other words, the singularities b_i **repel** each other, but the boundary condition on ∂G produces a **confinement effect**. In particular W achieved its minimum on G^d and every minimizing configuration consists of d **distinct points** in G^d **(not** $\bar{G}^d$**)**.

The location of the points (a_i) in Theorem 0.1 is governed by W through the following:

Theorem 0.2. *Let (a_i) be as in Theorem 0.1. Then (a_i) is a minimizer for W on G^d.*

The expression W comes up naturally in the following computation. Given any configuration $b = (b_1, b_2, \ldots, b_d)$ of distinct points in G, let $G_\rho = G \setminus \bigcup_i B(b_i, \rho)$. Consider the class

$$\mathcal{E}_\rho = \left\{ v \in H^1(G_\rho; S^1) \,\middle|\, \begin{matrix} v = g \text{ on } \partial G \quad \text{and} \\ \deg(v, \partial B(b_i, \rho)) = 1 \quad \forall i \end{matrix} \right\}. \tag{13}$$

One proves (see Theorem I.2) that there exists a unique minimizer u_ρ for the problem

(14) $$\operatorname*{Min}_{u\in\mathcal{E}_\rho}\int_{G_\rho}|\nabla u|^2$$

and that (see Theorem I.7) the following expansion holds:

(15) $$\frac{1}{2}\int_{G_\rho}|\nabla u_\rho|^2=\pi\, d|\log\,\rho|+W(b)+O(\rho)\quad\text{as } \rho\to 0.$$

In other words, W is what remains in the energy after the singular "**core energy**" $\pi\, d|\log\,\rho|$ has been removed. (The idea of removing an infinite core energy is common in physics; see e.g., M. Kléman [1]). Moreover, as $\rho\to 0$, u_ρ converges to some u_0 that has the following properties:

(16) $$u_0 \text{ is a smooth harmonic map in } G\setminus\bigcup_i\{b_i\}$$

(17) $$u_0=g\qquad\text{on } \partial G$$

(18) $$\left|u_0(z)-\beta_i\frac{(z-b_i)}{|z-b_i|}\right|\leq C|z-b_i|\qquad\text{as } z\to b_i,\ \forall i$$

for some complex numbers β_i with $|\beta_i|=1$ $\forall i$.

In fact, given any configuration $b\in G^d$ of distinct points, there is a unique u_0 satisfying (16), (17) and (18) (see Corollary I.1). We call this u_0 the **canonical harmonic map** associated to the configuration b.

There is an **explicit formula** for u_0 (see Corollary I.2):

(19) $$u_0(z)=e^{i\varphi(z)}\frac{(z-b_1)}{|z-b_1|}\frac{(z-b_2)}{|z-b_2|}\cdots\frac{(z-b_d)}{|z-b_d|}$$

where φ is the solution of the Dirichlet problem

(20) $$\begin{cases}\Delta\varphi=0 & \text{in } G\\ \varphi=\varphi_0 & \text{on } \partial G\end{cases}$$

and φ_0 is defined on ∂G by

(21) $$e^{i\varphi_0(z)}=g(z)\frac{|z-b_1|}{(z-b_1)}\frac{|z-b_2|}{(z-b_2)}\cdots\frac{|z-b_d|}{(z-b_d)}.$$

(Note that the right-hand side in (21) is a map from ∂G into S^1 of degree zero so that φ_0 is well defined as a single-valued smooth function.)

For a **general configuration** b estimate (18) **cannot** be improved. However, for the **special** configuration as described in Theorem 0.1 we have the better estimate (11). That property, which may be written as

$$\nabla\left(u_\star(x)\frac{|x-a_i|}{(x-a_i)}\right)(a_i)=0 \quad \forall i, \tag{22}$$

is related to the fact that $a=(a_1,a_2,\dots,a_d)$ is a critical point of W on G^d. It is extremely useful in localizing the singularities of $u_\star$ (see Section VIII.4).

The role of condition (22) has been strongly emphasized (in the case of a single singularity) by J. Neu[1] and by P. Fife and L. Peletier [1]. They show that (22) must be satisfied in order to be able to carry out a matched asymptotic expansion argument for (3).

Equation (22) also bears some resemblance with the results concerning the location of the blow-up points for the problem

$$-\Delta u = u^{p-\varepsilon} \qquad \text{or } \; -\Delta u = u^p+\varepsilon u \quad \text{in } \Omega\subset\mathbb{R}^n$$

with critical exponent $p=(n+2)/(n-2)$. There, the blow-up points a satisfy

$$\nabla H(a)=0$$

where H is the regular part of the Green's functions (see H. Brezis and L. Peletier [1] and O. Rey [1], [2]).

To complete the description of $u_\star$ we have:

Theorem 0.3. *Let (a_i) and $u_\star$ be as in Theorem 0.1. Then $u_\star$ is the canonical harmonic map associated to the configuration*

$$a=(a_1,a_2,\dots,a_d).$$

Conclusion: In general, W may have several minima. However, once the location of a_i is known, then $u_\star$ is completely determined. In some important cases W has a unique minimizer that can be identified explicitly; for example when $G=B_1$ and $g(x)=x$:

Theorem 0.4. *Assume $G=B_1$ and $g(x)=x$. Let u_ε be a minimizer for (1), then, $\forall x\neq 0$,*

$$u_\varepsilon(x)\to u_\star(x)=\frac{x}{|x|} \quad \textit{as } \varepsilon\to 0.$$

This answers Question 2 above.

Theorem 0.4 can be viewed as the 2-dimensional analogue of a result of H. Brezis, J. M. Coron and E. Lieb [1], which asserts that the unique minimizer of the problem

$$\operatorname*{Min}_{u\in H^1_g(B^3;S^2)} \int_{B^3} |\nabla u|^2 \quad \text{with } g(x) = x$$

is $u(x) = x/|x|$. More generally, F.H. Lin [1] has obtained the same conclusion for the problem

$$\operatorname*{Min}_{u\in H^1_g(B^n;S^{n-1})} \int_{B^n} |\nabla u|^2 \quad \text{for any } n \geq 3.$$

Next, we study the zeroes of u_ε. Let us recall some earlier works on that question. It has been proved by C. Elliott, H. Matano and T. Qi [1] that (for every $\varepsilon > 0$) the zeroes of any minimizer u_ε of (2) are isolated. P. Bauman, N. Carlson and D. Phillips [1] have shown, in particular, that if $G = B_1$ and $\deg(g, \partial G) = 1$ with $g(\theta)$ strictly increasing then (for every $\varepsilon > 0$) there is a unique zero of any minimizer u_ε of (2).

Our main result concerning the zeroes of u_ε is the following:

Theorem 0.5. *Let G be a starshaped domain and let $d = \deg(g, \partial G)$. Then, for $\varepsilon < \varepsilon_0$ depending only on g and G, u_ε has exactly d zeroes of degree* $+1$.

Remark 0.1. If $d \geq 2$ we give an example in Section VIII.5 showing that the conclusion of Theorem 0.5 fails when ε is large. The following happens: when ε is large u_ε has a single zero of degree d and, as $\varepsilon \to 0$, this zero splits into d zeroes of degree $+1$.

Finally we analyze the behavior as $\varepsilon \to 0$ of solutions v_ε of the Ginzburg-Landau equation (3), which **need not be minimizers** of E_ε. We prove that some of the results presented above for minimizers still hold for solutions of (3). In particular, v_{ε_n} converges to some limit $v_\star$ in $C^k_{\text{loc}}(G\backslash\bigcup_j\{a_j\})$ where $\{a_j\}$ is a finite set. However, by contrast with the previous situation, we have no information about $\text{card}(\bigcup_j\{a_j\})$ and $\deg(v_\star, a_j)$ need not be $+1$. More precisely, we have

Theorem 0.6. *Assume G is starshaped. Then there exist a subsequence $\varepsilon_n \to 0$, k points $a_1, a_2, \ldots, a_k$ in G and a smooth harmonic map $v_\star : \overline{G} \setminus \bigcup_j \{a_j\} \to S^1$ with $v_\star = g$ on ∂G such that*

$$v_{\varepsilon_n} \to v_\star \text{ in } C^\ell_{\text{loc}}(G \setminus \cup_j\{a_j\})\ \forall \ell \text{ and in } C^{1,\alpha}(\overline{G} \setminus \cup_j\{a_j\}) \quad \forall \alpha < 1.$$

Moreover, there exist integers $d_1, d_2, \ldots, d_k \in \mathbb{Z} \setminus \{0\}$ *and a smooth harmonic function* $\varphi : \overline{G} \to \mathbb{R}$ *such that*

$$v_\star(z) = e^{i\varphi(z)} \frac{(z-a_1)^{d_1}}{|z-a_1|^{d_1}} \cdots \frac{(z-a_k)^{d_k}}{|z-a_k|^{d_k}}.$$

In addition, we have

$$\nabla \left(v_\star(z) \frac{|z-a_j|^{d_j}}{(z-a_j)^{d_j}} \right)(a_j) = 0 \quad \forall j,$$

which expresses that (a_j, d_j) *is a critical point of some appropriate renormalized energy* W.

Remark 0.2. We emphasize that k need not be equal to d. However there is a bound for k in terms of g and G, and similarly for $\sum_j |d_j|$. We also emphasize that Theorem 0.6 is of interest even in the case where $d = \deg(g, \partial\Omega) = 0$ (we recall that the result of F. Bethuel, H. Brezis and F. Hélein [2] concerns only the analysis, as $\varepsilon \to 0$, of **minimizers** of E_ε when $d = 0$).

Analogies in physics. The results discussed in this book present striking analogies to numerous theoretical and experimental discoveries in the area of superconductors and superfluids over the past 40 years. Functionals of the form $E_\varepsilon(u)$ were originally introduced by V. Ginzburg and L. Landau [1] in the study of phase transition problems occurring in superconductivity; similar models are also used in superfluids such as helium II (see V. Ginzburg and L. Pitaevskii [1]) and in XY-magnetism. There is a considerable amount of literature on this huge subject; some of the standard references are: P. G. DeGennes [1], R. Donnelly [1], J. Kosterlitz and D. Thouless [1], D. Nelson [1], P. Nozières and D. Pines [1], R. Parks [1], D. Saint-James, G. Sarma and E. J. Thomas [1], D. Tilley and J. Tilley [1], M. Tinkham [1]. The unknown u represents a complex order parameter (i.e., with two degrees of freedom). In the physics literature u — often denoted ψ — is called a **condensate wave function** or a Higgs field. The parameter ε, which has the **dimension of a length**, depends on the material and its temperature. In the physics literature it is called the **(Ginzburg-Landau) coherence length** (or healing length or **core radius**) and is often denoted by $\xi = \xi(T)$. For temperatures $T < T_c$ (the critical temperature) with T not too close to T_c, $\xi(T)$ is **extremely small**, typically of the order of some hundreds of angstroms in superconductors, and of the order of a few angstroms in superfluids. Hence, it is of interest to study the asymptotics as $\varepsilon \to 0$, even though the limiting problem (at

$\varepsilon = 0$) has no physical meaning. Note that $\xi(T)$ plays the role of a **characteristic length**: the values of $|\psi(x)|$ may vary significantly at two points x_1, x_2 whose distance $|x_1 - x_2|$ is of the order of $\xi(T)$.

[**Warning:** Instead of equation (3), i.e.,

$$-\Delta\psi = \frac{1}{\xi^2}\psi(1 - |\psi|^2) \quad \text{in } G$$

some authors work with

$$-\Delta\hat{\psi} = \kappa^2\hat{\psi}(1 - |\hat{\psi}|^2) \text{ in } \hat{G} = \frac{1}{\lambda}G.$$

This amounts to a dilation in the space variables: $\hat{\psi}(x) = \psi(\lambda x)$ where $\lambda = \lambda(T)$ is the (London) penetration depth (another constant, having the dimension of a length, which also depends on the material and the temperature) and $\kappa = \lambda/\xi$ is the Ginzburg-Landau parameter, which is dimensionless (and need not be very large)].

In superconductors $|\psi|^2$ is proportional to the density of superconducting electrons (i.e., $|\psi| \simeq 1$ corresponds to the superconducting state and $|\psi| \simeq 0$ corresponds to the normal state). In superfluids $|\psi|^2$ is proportional to the density of the superfluid. If one writes $\psi = |\psi|e^{iS}$ where S is the real-valued phase, then its gradient ∇S is proportional to the velocity of the supercurrents or the superfluid.

Our analysis deals with the study of a 2-dimensional cross-section of a cylinder (a solid torus would also be of interest). In physical situations the Dirichlet condition is not realistic. However, it is striking to see that the degree $d = \deg(g, \partial G)$ of the boundary condition creates the same "**quantized vortices**" as a magnetic field in type-II superconductors or as an angular rotation in superfluids.

Vortex lines (or filaments) are produced in helium II by cooling a rotating bucket around the z-axis at constant angular velocity Ω. The vortex lines are parallel to the z-axis and they have a core radius of order ξ. The theory was initiated by L. Onsager [1] and R. Feynman [1]; the first experimental evidence came in the work of W. Vinen [1]. At high velocity Ω the number of vortices is proportional to Ω and they arrange themselves in a regular pattern—a triangular array; see e.g., the numerous pictures in the book of R. Donnelly [1]. Near the cores of the vortices the superfluid density $|\psi|^2$ is almost zero; away from the cores, $|\psi|^2 \simeq 1$.

Vortex lines are induced in type-II superconductors by applying a magnetic field H. For large H the number of vortices is proportional to H. This is the so-called **mixed state** (or vortex state) characterized by the **coexistence of two phases**: near the cores of the vortices $|\psi| \simeq 0$, i.e., normal state; away from the cores of the vortices $|\psi| \simeq 1$, i.e., superconducting state. Again, vortices have a core radius of order ξ. As H increases the vortices arrange themselves in a regular pattern, the **Abrikosov lattice** (predicted on a theoretical basis by A. Abrikosov [1]).

In the cross-section picture ψ seems to have basically the same behavior as our u_ε for ε small with vortices located around (and close to) the zeroes of u_ε (or in the $\varepsilon \to 0$ limit at the singularities of $u_\star$). The conclusion of Theorem 0.1 is consistent with the observation that all vortices have the same circulation +1 (in our language degree +1). Moreover, our analysis yields cores of radius ε that correspond to the physical cores of size ξ.

Remark 0.3. The conclusion of Theorem 0.1 bears some resemblance to earlier results motivated by the theory of nematic liquid crystals (see H. Brezis [1], [2], H. Brezis, J. M. Coron and E. Lieb [1], R. Hardt and F. H. Lin [1], D. Kinderlehrer [1]). Let $G \subset \mathbb{R}^3$ be a smooth bounded domain and let $g : \partial G \to S^2$ be a (smooth) boundary condition. We now consider maps $u : G \to \mathbb{R}^3$ (not $\mathbb{R}^2$); in the theory of liquid crystals u corresponds to the "director", which is an order parameter describing the orientation of the optical axis of the (rod-like) molecules. As above, set

$$H^1_g(G;\mathbb{R}^3) = \{u \in H^1(G;\mathbb{R}^3);\ u = g \text{ on } \partial G\}$$

and

$$H^1_g(G;S^2) = \{u \in H^1_g(G;\mathbb{R}^3); |u(x)| = 1 \text{ a.e.}\}.$$

In contrast with the above situation, here,

$$H^1_g(G;S^2) \neq \emptyset \qquad \forall g;$$

for example, if G is the unit ball then

$$u(x) = g(x/|x|) \in H^1_g(G;S^2)$$

since, now,

$$\int_G \left|\nabla\left(\frac{x}{|x|}\right)\right|^2 < \infty.$$

Consider the same energy E_ε as above on $H^1_g(G;\ \mathbb{R}^3)$ and let u_ε be a minimizer of E_ε. It is easy to prove that $u_{\varepsilon_n} \to u_0$ in H^1, where u_0 is a minimizing harmonic map from G into S^2 i.e., u_0 is a minimizer of $\int_G |\nabla u|^2$ on $H^1_g(G;S^2)$. An important result of R. Schoen and K. Uhlenbeck [1], [2] asserts that u_0 has a finite number of singularities. But, it is extremely difficult to estimate the number of singularities (in terms of g and G); see however an interesting contribution in that direction by F. Almgren and E. Lieb [1]. Moreover, H. Brezis, J. M. Coron and E. Lieb [1] have proved that near a singular point a, u_0 has a simple behaviour:

$$u_0(x) \cong \pm R\left(\frac{x-a}{|x-a|}\right) \quad \text{as } x \to a$$

for some rotation R. As a consequence, the degree of each singularity is $+1$ or -1. Singularities of degree $+1$ and -1 **may coexist** (see e.g., R. Hardt and F. H. Lin [1], H. Brezis [1]). This is a striking difference with our Theorem 0.1 which **excludes such coexistence.** Roughly speaking, this phenomenon is related to the fact that in the 2-d problem singularities have infinite energy while in the 3-d problem they have finite energy (see the precise analysis of Chapters V and VI).

Another striking difference is that nothing is known about the location of singularities in the 3-d problem while Theorem 0.2 (and (11)) provides a very precise information about the location of singularities of $u_\star$.

Theorem 0.1 is also in agreement with the numerical simulations of Carlson and Miller presented in J. Neu [1]:

(i) a pair of singularities of degrees $+1$ and -1 is **unstable** and tends to **coalesce**;

(ii) a singularity of degree $d > 1$ is also **unstable** and tends to **split** into d singularities of degree $+1$.

Remark 0.4. The results presented here differ considerably from the conclusions obtained by numerous authors when u is a **scalar function,** i.e., $u : G \to \mathbb{R}$; the motivation there comes from the Van der Waals and Cahn-Hilliard theory of phase transition (see M. Gurtin [1], L. Modica [1], P. Sternberg [1], R. Kohn and P. Sternberg [1], E. DeGiorgi [2]). In the scalar case $u_\varepsilon \to u_\star$ which takes only the values $+1$ and -1; the sets $[u_\star = +1]$ and $[u_\star = -1]$ are separated by an interface S having minimal area. The phase transition region $[-1+\delta < u_\varepsilon < 1-\delta]$ consists of a thin layer enclosing S (see Figure 1). By contrast, in our situation the phase transition region $[|u_\varepsilon| < 1-\delta]$ occurs in small neighborhoods of point singularities (see Figure 2).

The difference in the analysis stems from the fact that, in the scalar case, the potential $(|u|^2-1)^2$ is a **two-wells potential** while in the complex case the same potential has an $\mathbf{S}^1$**-well.** In the scalar case it is easy to pass to the limit a.e. once the natural estimate $\int_G |\nabla\varphi(u_\varepsilon)| \le C$, for some appropriate

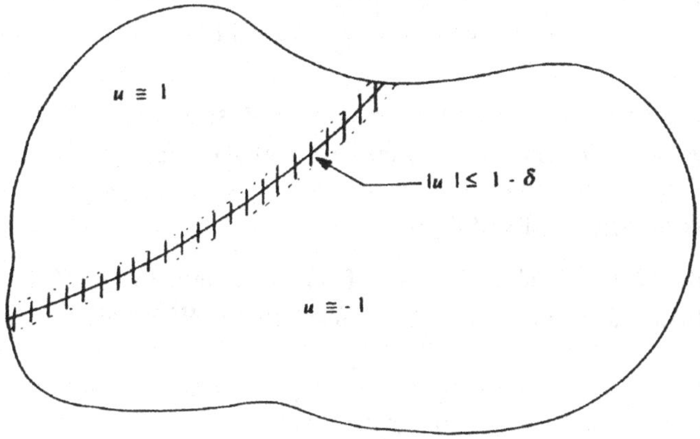

FIGURE 1.

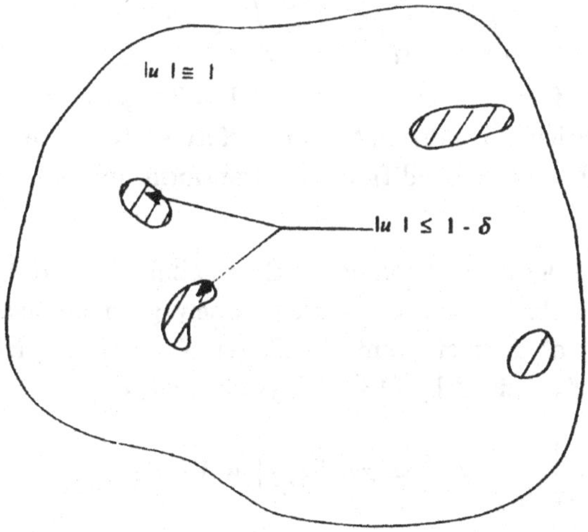

FIGURE 2.

function φ, has been established (such an estimate is obtained by a simple, although ingenious device; see L. Modica [1]). In our situation, it turns out that $\int_G |\nabla u_\varepsilon|$ is bounded (and even $\int_G |\nabla u_\varepsilon|^p$ for any $p < 2$), but the argument is much more involved (see Chapter X).

A final comment:

Remark 0.5. Most of the results presented here hold (with identical proofs) if the energy $E_\varepsilon(u)$ is replaced by

$$\frac{1}{2}\int_G |\nabla u|^2 + \frac{1}{2}\int_G F(1-|u|^2)$$

where F is a (smooth) convex function on $\mathbb{R}$ such that $F(0) = 0$ and $F(t) > 0, \ \forall\, t > 0$; see however, Open Problem 2 in Chapter XI.

The book is organized as follows:

In Chapter I we consider maps u from a domain $\Omega \subset \mathbb{R}^2$ with holes (possibly shrinking to points) with values into S^1. We study

$$\operatorname*{Min}_{u:\Omega\to S^1} \int_\Omega |\nabla u|^2$$

under various boundary conditions on $\partial\Omega$: Dirichlet condition, prescribed degree on the boundary of the holes, or a combination of both. The main property is that

$$\operatorname*{Min}_{u:\Omega\to S^1} \int_\Omega |\nabla u|^2 = \int_\Omega |\nabla \Phi|^2$$

where Φ is a **scalar function** that satisfies $\Delta\Phi = 0$ together with boundary conditions. The Dirichlet condition for u transforms into a Neumann condition for Φ, while the degree condition on u transforms into a Dirichlet-type condition for Φ.

We also study in detail the behavior of the minimizer u and its energy as the size of the holes shrinks to zero. More precisely, given any configuration $b = (b_1, b_2, \ldots, b_d)$ of distinct points in G, set $G_\rho = G \setminus \bigcup_i B(b_i,\, \rho)$ and consider the class $\mathcal{E}_\rho$ defined by (13). We prove that, as $\rho \to 0$,

$$\operatorname*{Min}_{u\in\mathcal{E}_\rho} \frac{1}{2}\int_{G_\rho} |\nabla u|^2 = \pi\, d|\log \rho| + W(b) + O(\rho)$$

where $W(b)$ is the renormalized energy defined by (12).

In Chapter II we estimate $\int_\Omega |\nabla\Phi|^2$ from below, where $\Omega = G \setminus \bigcup_i B(b_i,\, \rho)$, under a Dirichlet-type condition for Φ. The main result (Theorem II.1) is somewhat technical, but it plays a crucial role in the proof of Theorem 0.1. The underlying idea is that a singularity of degree m has a "core energy" of the order

$$\pi\, m^2 |\log \rho|.$$

This leads to the important observation that **one singularity of degree $m > 1$ carries more energy than m singularities of degree 1.**

In Chapter III we start the study of Problem (2) and we derive some elementary but basic estimates. The main estimates are the following:

$$E_\varepsilon(u_\varepsilon) \le \pi d|\log \varepsilon| + C. \tag{23}$$

Under the additional assumption that G is starshaped we obtain a better estimate for the second term in $E_\varepsilon(u_\varepsilon)$, namely, we have

$$\frac{1}{\varepsilon^2}\int_G (|u_\varepsilon|^2 - 1)^2 \le C^* \tag{24}$$

where the constants C and C^* depend only on G and g.

Estimate (24) plays a fundamental role in our analysis. In particular, it is used to prove that $|u_\varepsilon| \ge 1/2$ outside a finite number of discs of size ε (see (25) below).

The proof of Theorem 0.1 (except for estimate (11)) runs through Chapters IV, V and VI. We briefly describe the strategy of the proof.

We first use (24) together with a simple covering argument to isolate, for every ε, a finite number of "bad discs" $B(x_i^\varepsilon, \lambda\varepsilon), i \in J_\varepsilon$, with $x_i^\varepsilon \in G$ such that

$$|u_\varepsilon| \ge \frac{1}{2} \quad \text{in } G \setminus \cup_{i\in J_\varepsilon} B(x_i^\varepsilon, \lambda\varepsilon) \tag{25}$$

and

$$\operatorname{card} J_\varepsilon \le N, \tag{26}$$

where the constant $\lambda > 0$ and the integer N depend only on G and g (they are independent of ε).

As $\varepsilon \to 0$ (along a subsequence) the points x_i^ε converge to some points denoted $(a_j)_{j\in J}$ with $a_j \in \overline{G}$ and card $J \le N$. The points (a_j) are the natural candidates for being the singularities of lim u_{ε_n} (assuming such a limit exists — which we don't know yet!).

A central part of the proof consists in showing that, for any fixed $\eta > 0$, if we set

$$G_\eta = G \setminus \cup_{j\in J} B(a_j, \eta),$$

then

$$\int_{G_\eta} |\nabla u_{\varepsilon_n}|^2 \le C(\eta) \tag{27}$$

where $C(\eta)$ depends only on η, G and g (but not on ε_n). This is the content of Theorem V.1.

For this purpose, we establish, in Chapter V, **lower bounds** for

$$\int_{B(a_j,\eta)} |\nabla u_{\varepsilon_n}|^2.$$

Set

$$\kappa_j = \deg(u_{\varepsilon_n}, \partial B(a_j,\eta)). \tag{28}$$

Note that this degree makes sense since

$$|u_{\varepsilon_n}| \geq \frac{1}{2} \text{ on } \partial B(a_j, \eta) \text{ by (25).}$$

(In principle $\kappa_j = \kappa_j^n$ depends on n, but we provide a bound for κ_j^n independent of n so that, by passing to a further subsequence, we may always assume that κ_j^n is independent of n. Another difficulty stems from the fact that a_j may lie on ∂G; to get around this difficulty it is convenient to enlarge a little bit G, say by a domain G', and to extend u_ε by a fixed map $\overline{G}$ on $G' \setminus G$ with $|\overline{G}| = 1$ on $G' \setminus G$).

Roughly speaking, we prove that, for every $\eta > 0$,

$$\frac{1}{2}\int_{B(a_j,\eta)} |\nabla u_{\varepsilon_n}|^2 \geq \pi\kappa_j^2 |\log \varepsilon_n| - C(\eta), \quad \forall j, \forall n \geq N(\eta). \tag{29}$$

[The actual estimates presented in Chapters V and VI are technically more complicated, but (29) represents a good heuristic way of understanding them].

Combining (29) with the basic estimate (23) we see that

$$\sum_{j\in J} \kappa_j^2 \leq d. \tag{30}$$

On the other hand, we clearly have

$$\sum_{j\in J} \kappa_j = d. \tag{31}$$

From (30) and (31) we deduce that

$$\sum_{j\in J} (\kappa_j^2 - \kappa_j) \leq 0.$$

Since we obviously have $\kappa_j^2 - \kappa_j \geq 0 \quad \forall j$, it follows that

$$\kappa_j^2 - \kappa_j = 0 \ \forall j, \ \text{ i.e., } \kappa_j \in \{0,1\} \quad \forall j.$$

Using the results of F. Bethuel, H. Brezis and F. Hélein [2], we are able to exclude the possibility that $\kappa_j = 0$. We are thus left with

$$\kappa_j = 1 \qquad \forall j \tag{32}$$

and

$$\text{card } J = d. \tag{33}$$

If one of the points a_j, say a_1, belongs to ∂G we may improve (29). Instead of (29) (with $\kappa_1 = 1$) we now have

$$\frac{1}{2}\int_{B(a_1,\eta)} |\nabla u_{\varepsilon_n}|^2 \geq 2\pi |\log \varepsilon_n| - C. \tag{34}$$

Repeating the same argument as above we are led to

$$(2-1) + \sum_{j\neq 1}(\kappa_j^2 - \kappa_j) \leq 0$$

which is impossible. Hence $a_j \in G \quad \forall j$.

We may now derive (27) very easily since

$$\begin{aligned}\frac{1}{2}\int_{G_\eta} |\nabla u_{\varepsilon_n}|^2 = \frac{1}{2}\int_G |\nabla u_{\varepsilon_n}|^2 - \frac{1}{2}\sum_j \int_{B(a_j,\eta)} |\nabla u_{\varepsilon_n}|^2 \\ \leq \pi d|\log \varepsilon_n| - \pi d |\log \varepsilon_n| + C = C\end{aligned}$$

by (23) and (29).

Once (27) is established, we conclude by a standard diagonal argument that, for a subsequence of ε_n, we have

$$u_{\varepsilon_n} \to u_\star \text{ a.e., in } G.$$

The convergence in stronger norms announced in Theorem 0.1 follows from (27) and the results of our earlier work, F. Bethuel, H. Brezis and F. Hélein [2] (see Chapter VI). This yields a smooth harmonic map $u_\star$ on $\overline{G}\setminus \bigcup\{a_j\}$ with precisely d singularities in G, each one of degree $+1$.

Theorem 0.2 and estimate (11) in Theorem 0.1 are proved in Chapters VII and VIII. In Chapter VII we use the stationarity of u_ε with respect to deformations of u_ε induced by the group of diffeomorphisms of G (in the same spirit as Pohozaev-type identities). This leads to a precise description of the Hopf differential

$$\omega = \left|\frac{\partial u_\star}{\partial x_1}\right|^2 - \left|\frac{\partial u_\star}{\partial x_2}\right|^2 - 2i\,\frac{\partial u_\star}{\partial x_1}\cdot\frac{\partial u_\star}{\partial x_2}$$

of the limiting map $u_\star$ near its singularities a_j. (**Warning:** Here the dot product refers to the scalar product, not complex multiplication). These informations yield a good control of the behavior of $u_\star$ near a_j, which, combined with the results of Chapter I, allows us to identify $u_\star$ with the canonical harmonic map defined above (see (19)).

In Chapter VIII we prove that the configuration (a_j) minimizes the renormalized energy W using appropriate comparison functions and the strong convergence of u_{ε_n} in C^k norms away from the singularities. We also discuss the relationship between the fact that $a = (a_1, a_2, \ldots, a_d)$ is a critical point of W and property (22) (or equivalently (11)). In Sections VIII.4 and VIII.5 we study two specific examples: $G = B_1$ with $g(\theta) = e^{i\theta}$ and $g(\theta) = e^{di\theta}$. In particular, we prove Theorem 0.4.

Section IX.1 is devoted to the proof of Theorem 0.5. In Section IX.2 we prove that (see Theorem IX.3)

$$\lim_{\varepsilon\to 0}\{E_\varepsilon(u_\varepsilon) - \pi\, d|\log\varepsilon|\} = \operatorname*{Min}_{b\in G^d} W(b) + d\gamma \tag{35}$$

where γ is some universal constant.

In view of (35) it would be interesting to study the minimization problem (2) in the framework of the Γ-convergence theory introduced by E. DeGiorgi (see e.g., E. DeGiorgi [1], E. DeGiorgi and T. Franzoni [1], L. Modica and S. Mortola [1]). Our results suggest that the functionals

$$F_\varepsilon(u) = E_\varepsilon(u) - \pi d|\log\varepsilon|$$

Γ-converge to some kind of renormalized energy

$$F_0(u) = \liminf_{\rho\to 0}\left\{\frac{1}{2}\int_{G_\rho}|\nabla u|^2 - \pi d|\log\rho|\right\}$$

where $b_j \in G$, $j = 1, 2, \ldots, d$, $u \in H^1_{\mathrm{loc}}(G\setminus\bigcup_j\{b_j\}; S^1)$, $u = g$ on ∂G and $G_\rho = G\setminus\bigcup_j B(b_j,\rho)$.

Finally, in Chapter X, we prove Theorem 0.6. The starting point is again a covering argument that isolates bad discs as in (25)–(26). But, next we use a different strategy: we prove that

$$\int_G |\nabla u_\varepsilon|^p \leq C_p, \quad \forall p < 2$$

where C_p depends only on g, G and p. This involves the use of estimates for linear elliptic equations in divergence form à la Stampacchia [1].

We conclude, in Chapter XI, with a list of numerous open problems related to (2) and (3).

Some of the main results were announced without proofs in F. Bethuel, H. Brezis and F. Hélein [1],[3].

CHAPTER I

Energy estimates for S^1-valued maps

I.1. An auxiliary linear problem

Let G be a smooth, bounded and simply connected domain in $\mathbb{R}^2$, and let ω_i, for $i = 1, \ldots, n$, be open, smooth and simply connected subsets of G, with $\bar{\omega}_i \subset G$ and $\bar{\omega}_i \cap \bar{\omega}_j = \emptyset$ for $i \neq j$. Let $\Omega = G \setminus \bigcup_{i=1}^{n} \bar{\omega}_i$. Consider the class of maps

$$(1) \qquad \mathcal{E} = \left\{ v \in H^1(\Omega; S^1) \,\middle|\, \begin{array}{ll} \deg(v, \partial G) = d & \text{and} \\ \deg(v, \partial\omega_i) = d_i & \text{for } i = 1, 2, \ldots, n \end{array} \right\}$$

where $d_i \in \mathbb{Z}$ are given and $d = \sum_{i=1}^{n} d_i$.

We study the minimization problem

$$(2) \qquad E = \operatorname*{Inf}_{v \in \mathcal{E}} \int_\Omega |\nabla v|^2.$$

At this stage, it is not clear that the infimum is achieved since the function $v \mapsto \deg(v, \partial\omega_i)$ is not continuous under weak $H^{1/2}(\partial\omega_i)$ convergence; the existence of a minimizer will be derived as a consequence of the discussion below.

The value of E is related to the solution Φ of the following linear problem:

$$(3) \qquad \begin{cases} \Delta\Phi = 0 & \text{in } \Omega, \\ \Phi = \text{Const.} = C_i & \text{on } \partial\omega_i, \quad i = 1, 2, \ldots, n, \\ \Phi = 0 & \text{on } \partial G, \\ \displaystyle\int_{\partial\omega_i} \frac{\partial\Phi}{\partial\nu} = 2\pi d_i & i = 1, 2, \ldots, n, \end{cases}$$

where ν is the outward normal to ω_i and also the outward normal to G. Here the C_i's are not given but are unkown constants that are part of the problem. It is well known (see R. Temam [1], H. Berestycki and H. Brezis

[1], [2]) that problem (3) has a unique solution. Moreover, Φ is obtained by minimizing

(4) $$F(\varphi) = \frac{1}{2}\int_\Omega |\nabla\varphi|^2 + 2\pi \sum_{i=1}^{n} d_i\, \varphi_{|\partial\omega_i}$$

in the class

$$V = \{\varphi \in H^1(\Omega;\mathbb{R}); \varphi = 0 \text{ on } \partial G, \varphi = \text{Const.} = \varphi_{|\partial\omega_i} \text{ on each } \partial\omega_i\}.$$

Our main result is the following.

Theorem I.1. *The infimum in (2) is achieved by a map that is unique up to a phase, i.e., if u_1 and u_2 are two minimizers, then $u_1 = \alpha u_2$ where $\alpha \in \mathbb{C}$ with $|\alpha| = 1$. Moreover*

(5) $$E = \int_\Omega |\nabla\Phi|^2.$$

We split the proof into two steps.

Step 1: For any v in $\mathcal{E}$,

$$\int_\Omega |\nabla v|^2 \geq \int_\Omega |\nabla\Phi|^2.$$

Proof of Step 1. Since v takes its values in S^1, we have

$$v_{x_1} \times v_{x_2} = 0;$$

hence

(6) $$\frac{\partial}{\partial x_1}(v \times v_{x_2}) + \frac{\partial}{\partial x_2}(-v \times v_{x_1}) = 0.$$

Set

$$D = (-v \times v_{x_2} + \Phi_{x_1},\ v \times v_{x_1} + \Phi_{x_2}).$$

From (3) and (6) we have

(7) $$\operatorname{div} D = 0.$$

On the other hand we also have

(8) $$\int_{\partial\omega_i} D \cdot \nu = 0 \quad \text{for each } i.$$

Indeed,

$$D \cdot \nu = -(v \times v_\tau) + \frac{\partial \Phi}{\partial \nu},$$

where τ is the unit tangent vector to $\partial\omega_i$ such that (ν, τ) is direct (see Figure 3). Then (8) follows from the fact that

$$\int_{\partial\omega_i} v \times v_\tau = 2\pi \ \deg(v, \partial\omega_i) = 2\pi \ d_i$$

and

$$\int_{\partial\omega_i} \frac{\partial \Phi}{\partial \nu} = 2\pi \ d_i.$$

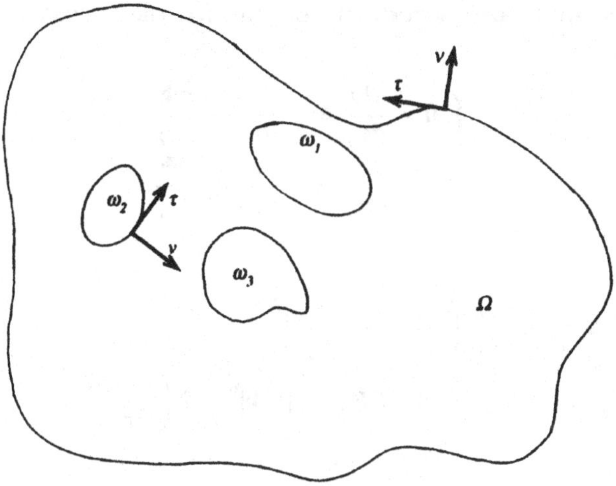

FIGURE 3.

We now use the following standard lemma.

Lemma I.1. *Let Ω be any smooth open domain in $\mathbb{R}^2$ (not necessarily simply connected) and D be a vector field on Ω such that*

$$div\ D = 0,$$

and

$$\int_{\Gamma_i} D \cdot \nu = 0$$

for each connected component Γ_i of $\partial\Omega$. Then there exists a function H on Ω such that

$$D = \left(\frac{\partial H}{\partial x_2}, -\frac{\partial H}{\partial x_1}\right).$$

Sketch of the proof. If Ω is simply connected, this is well known (Poincaré's lemma). In the case of a general domain Ω (which we may always assume to be connected) we solve in each domain ω_i enclosed by Γ_i the problem

$$\begin{cases} \Delta w_i = 0 & \text{in } \omega_i \\ \dfrac{\partial w_i}{\partial \nu} = D \cdot \nu & \text{on } \partial\omega_i. \end{cases}$$

The vector field

$$\tilde{D} = \begin{cases} D & \text{in } \Omega, \\ \nabla w_i & \text{in } \omega_i\,,\ i = 1, 2, \ldots, n, \end{cases}$$

satisfies $\operatorname{div} \tilde{D} = 0$ in $G = \Omega \bigcup \left(\bigcup_{i=1}^{n} \bar{\omega}_i \right)$, which is simply connected.

Proof of Step 1 completed. By Lemma I.1, there exists a function H such that

$$\text{(9)} \qquad \begin{cases} v \times \dfrac{\partial v}{\partial x_1} = -\dfrac{\partial H}{\partial x_1} - \dfrac{\partial \Phi}{\partial x_2} \\ v \times \dfrac{\partial v}{\partial x_2} = -\dfrac{\partial H}{\partial x_2} + \dfrac{\partial \Phi}{\partial x_1}. \end{cases}$$

Thus,

$$\text{(10)} \quad \begin{aligned} |\nabla v|^2 &= \left| v \times \frac{\partial v}{\partial x_1} \right|^2 + \left| v \times \frac{\partial v}{\partial x_2} \right|^2 \\ &= |\nabla H|^2 + |\nabla \Phi|^2 + 2 \left(\frac{\partial H}{\partial x_1} \frac{\partial \Phi}{\partial x_2} - \frac{\partial H}{\partial x_2} \frac{\partial \Phi}{\partial x_1} \right). \end{aligned}$$

Note that

$$\text{(11)} \quad \begin{aligned} \int_\Omega \frac{\partial H}{\partial x_1} \frac{\partial \Phi}{\partial x_2} - \frac{\partial H}{\partial x_2} \frac{\partial \Phi}{\partial x_1} &= \int_\Omega \operatorname{div} \left(H \frac{\partial \Phi}{\partial x_2}, -H \frac{\partial \Phi}{\partial x_1} \right) \\ &= \int_{\partial G} H \frac{\partial \Phi}{\partial \tau} - \sum_{i=1}^{n} \int_{\partial \omega_i} H \frac{\partial \Phi}{\partial \tau} = 0 \end{aligned}$$

since Φ is constant on each component of $\partial\Omega$. From (10) and (11) we conclude that

$$\text{(12)} \qquad \int_\Omega |\nabla v|^2 = \int_\Omega |\nabla H|^2 + \int_\Omega |\nabla \Phi|^2 \geq \int_\Omega |\nabla \Phi|^2.$$

Step 2: There exists some u in $\mathcal{E}$ such that

$$\int_\Omega |\nabla u|^2 = \int_\Omega |\nabla \Phi|^2.$$

Proof of Step 2. In view of (12) we shall try to find some u in $\mathcal{E}$ that satisfies

$$(13)\qquad \begin{cases} u \times \dfrac{\partial u}{\partial x_1} = -\dfrac{\partial \Phi}{\partial x_2} \\ u \times \dfrac{\partial u}{\partial x_2} = \dfrac{\partial \Phi}{\partial x_1}. \end{cases}$$

More generally, consider the problem on an arbitrary smooth connected domain Ω

$$(14)\qquad \begin{cases} u \times \dfrac{\partial u}{\partial x_1} = F_1 & \text{in } \Omega, \\ u \times \dfrac{\partial u}{\partial x_2} = F_2 & \text{in } \Omega. \end{cases}$$

This problem has a solution $u : \Omega \to S^1$ if and only if

$$(15)\qquad \frac{\partial F_1}{\partial x_2} = \frac{\partial F_2}{\partial x_1}$$

and

$$(16)\qquad \int_{\Gamma_i} F \cdot \tau \in 2\pi\, \mathbb{Z}$$

for each connected component Γ_i of $\partial\Omega$. Note that in our situation (15) holds since Φ is harmonic and (16) holds since $F \cdot \tau = \dfrac{\partial \Phi}{\partial \nu}$.

Indeed we may write locally $u = e^{i\psi}$, and then (14) becomes

$$\begin{cases} \dfrac{\partial \psi}{\partial x_1} = F_1 \\ \dfrac{\partial \psi}{\partial x_2} = F_2 \end{cases}$$

which has a local solution by (15). Moreover if u_1 and u_2 are two local solutions of (14) then $u_1 = \alpha u_2$, where α is a complex constant with $|\alpha| = 1$. The local solution ψ is continued globally by integrating over paths. The corresponding integrals may differ, but the difference belongs to $2\pi\mathbb{Z}$, and thus $u = e^{i\psi}$ is well defined up to a constant phase. Moreover u belongs to $\mathcal{E}$, since

$$2\pi \deg(u, \partial\omega_i) = \int_{\partial\omega_i} u \times u_\tau = \int_{\partial\omega_i} \frac{\partial \Phi}{\partial \nu} = 2\pi\, d_i.$$

I.2. Variants of Theorem I.1

With G, Ω, d_i and d as above, consider the class

$$\mathcal{E}_1 = \left\{ v \in H^1(\Omega; S^1) \middle| \begin{array}{l} v = g \text{ on } \partial G \quad \text{and} \\ \deg(v, \partial\omega_i) = d_i, i = 1, 2, \dots, n \end{array} \right\}$$

where $g : \partial G \to S^1$ is given such that

$$\deg(g, \partial G) = d = \sum_{i=1}^{n} d_i.$$

We study the minimization problem

$$E_1 = \operatorname*{Inf}_{v \in \mathcal{E}_1} \int_\Omega |\nabla v|^2. \tag{17}$$

As above, the value of E_1 is related to the solution of another linear problem.

$$\begin{cases} \Delta\Phi_1 = 0 & \text{in } \Omega, \\ \Phi_1 = \text{Const.} = C_i & \text{on } \partial\omega_i,\ i = 1, 2, \dots, n, \\ \displaystyle\int_{\partial\omega_i} \frac{\partial \Phi_1}{\partial \nu} = 2\pi\, d_i & i = 1, 2, \dots, n, \\ \dfrac{\partial \Phi_1}{\partial \nu} = g \times \dfrac{\partial g}{\partial \tau} & \text{on } \partial G. \end{cases} \tag{18}$$

This problem has a unique solution (up to an additive constant). Moreover Φ_1 is obtained by minimizing

$$\operatorname*{Inf}_{\varphi \in V_1} \left\{ \frac{1}{2} \int_\Omega |\nabla\varphi|^2 + 2\pi \sum_{i=1}^{n} d_i\, \varphi_{|\partial\omega_i} - \int_{\partial G} \varphi \left(g \times \frac{\partial g}{\partial \tau} \right) \right\}$$

where $V_1 = \{\varphi \in H^1(\Omega; \mathbb{R});\ \varphi = \text{Const.} = \varphi_{|\partial\omega_i} \text{ on each } \partial\omega_i\}$.

We have

Theorem I.2. *The infimum in (17) is achieved by a unique solution. Moreover*

$$E_1 = \int_\Omega |\nabla\Phi_1|^2.$$

Sketch of the proof. We follow the same argument as in Theorem I.1. The only difference in Step 1 is that $\int_{\partial G} H \dfrac{\partial \Phi_1}{\partial \tau} = 0$ since

$$\int_{\partial G} H \frac{\partial \Phi_1}{\partial \tau} = - \int_{\partial G} \Phi_1 \frac{\partial H}{\partial \tau}$$

and, by (9),

$$\frac{\partial H}{\partial \tau} = -\left(g \times \frac{\partial g}{\partial \tau}\right) + \frac{\partial \Phi_1}{\partial \nu} = 0 \quad \text{on } \partial G.$$

In Step 2, we first find some v in $\mathcal{E}$ that satisfies

$$\begin{cases} v \times \dfrac{\partial v}{\partial x_1} = -\dfrac{\partial \Phi_1}{\partial x_2} & \text{in } \Omega, \\ v \times \dfrac{\partial v}{\partial x_2} = \dfrac{\partial \Phi_1}{\partial x_1} & \text{in } \Omega. \end{cases}$$

For this v we have

$$v \times v_\tau = \frac{\partial \Phi_1}{\partial \nu} = g \times g_\tau \quad \text{on } \partial G$$

and it follows that $v = \alpha g$ on ∂G, where α is a complex constant with $|\alpha| = 1$. Then $u = \alpha^{-1} v$ satisfies all the required properties.

This last device works because we have prescribed a Dirichlet condition for u only on one component of $\partial\Omega$, namely ∂G. If a Dirichlet condition for u is prescribed on more than one component the situation becomes more complicated. For simplicity, let us consider the case where u satisfies a Dirichlet condition on all components of $\partial\Omega$, i.e., consider the class

$$\mathcal{E}_2 = \{v \in H^1(\Omega; S^1);\ v = g_i \text{ on } \partial\omega_i,\ i = 0, 1, 2, \dots, n\}$$

with $\partial\omega_0 = \partial G$ and (g_i) given such that

$$\deg(g_0, \partial G) = \sum_{i=1}^{n} \deg(g_i, \partial\omega_i).$$

Set

$$E_2 = \operatorname*{Min}_{v \in \mathcal{E}_2} \int_\Omega |\nabla v|^2. \tag{19}$$

The value of E_2 is related to the solution of the linear problem

$$\begin{cases} \Delta\Phi_2 = 0 & \text{in } \Omega, \\ \dfrac{\partial \Phi_2}{\partial \nu} = g_i \times \dfrac{\partial g_i}{\partial \tau} & \text{on } \partial\omega_i,\ i = 0, 1, 2, \dots, n. \end{cases} \tag{20}$$

Theorem I.3. *We have*

$$\int_\Omega |\nabla \Phi_2|^2 \le E_2 \le \int_\Omega |\nabla \Phi_2|^2 + 4\pi^2 \sum_{i=1}^n \operatorname{cap} \omega_i$$

where

$$\operatorname{cap} \omega_i = \int_\Omega |\nabla \psi_i|^2$$

and ψ_i is the solution of

$$\begin{cases} \Delta \psi_i = 0 & \text{in } \Omega, \\ \psi_i = 1 & \text{on } \partial\omega_i, \\ \psi_i = 0 & \text{on } \partial\Omega \backslash \partial\omega_i. \end{cases}$$

Proof.

Step 1: We prove that

$$E_2 \ge \int_\Omega |\nabla \Phi_2|^2.$$

We proceed as in the proof of Theorem I.1. Given v in $\mathcal{E}_2$, there exists a function H on Ω such that

$$\begin{cases} v \times \dfrac{\partial v}{\partial x_1} = -\dfrac{\partial H}{\partial x_1} - \dfrac{\partial \Phi_2}{\partial x_2} & \text{in } \Omega \\ v \times \dfrac{\partial v}{\partial x_2} = -\dfrac{\partial H}{\partial x_2} + \dfrac{\partial \Phi_2}{\partial x_1} & \text{in } \Omega. \end{cases}$$

By (20), $\dfrac{\partial H}{\partial \tau} = 0$ on $\partial\Omega$, and thus, as in the proof of Theorem I.1,

$$\int_\Omega |\nabla v|^2 = \int_\Omega |\nabla H|^2 + \int_\Omega |\nabla \Phi_2|^2 \ge \int_\Omega |\nabla \Phi_2|^2.$$

Step 2: We first find some $v : \Omega \to S^1$ that satisfies

$$\begin{cases} v \times \dfrac{\partial v}{\partial x_1} = -\dfrac{\partial \Phi_2}{\partial x_2} & \text{in } \Omega, \\ v \times \dfrac{\partial v}{\partial x_2} = \dfrac{\partial \Phi_2}{\partial x_1} & \text{in } \Omega. \end{cases}$$

This problem has a unique solution up to a phase. Moreover v satisfies

$$v \times \frac{\partial v}{\partial \tau} = \frac{\partial \Phi_2}{\partial \nu} = g_i \times \frac{\partial g_i}{\partial \tau} \quad \text{on } \partial\omega_i \text{ for } i = 0, 1, \ldots, n.$$

Choosing an appropriate phase we may always assume that

$$v = g_0 \quad \text{on } \partial\omega_0 = \partial G,$$

and then $v = e^{i\theta_j} g_j$ on $\partial\omega_j$ for $j = 1, \ldots, n$, for some θ_j in $[0, 2\pi]$.

Let ψ be the solution of

$$\begin{cases} \Delta\psi = 0 & \text{in } \Omega, \\ \psi = \theta_j & \text{on } \partial\omega_j, \text{ for } j = 1, 2, \ldots, n, \\ \psi = 0 & \text{on } \partial G. \end{cases}$$

Set $u = e^{-i\psi} v$ so that $u \in \mathcal{E}_2$ and

$$\begin{cases} u \times \dfrac{\partial u}{\partial x_1} = v \times \dfrac{\partial v}{\partial x_1} - \dfrac{\partial \psi}{\partial x_1} = -\dfrac{\partial \Phi_2}{\partial x_2} - \dfrac{\partial \psi}{\partial x_1} \\ u \times \dfrac{\partial u}{\partial x_2} = v \times \dfrac{\partial v}{\partial x_2} - \dfrac{\partial \psi}{\partial x_2} = \dfrac{\partial \Phi_2}{\partial x_1} - \dfrac{\partial \psi}{\partial x_2}. \end{cases}$$

Hence

$$\begin{aligned} |\nabla u|^2 &= |\nabla\psi|^2 + |\nabla\Phi_2|^2 - 2\left(\frac{\partial\Phi_2}{\partial x_1}\frac{\partial\psi}{\partial x_2} - \frac{\partial\Phi_2}{\partial x_2}\frac{\partial\psi}{\partial x_1}\right) \\ &= |\nabla\psi|^2 + |\nabla\Phi_2|^2 + 2\operatorname{div}\left[\psi\frac{\partial\Phi_2}{\partial x_2}, -\psi\frac{\partial\Phi_2}{\partial x_1}\right] \end{aligned}$$

and therefore, as in (11), we obtain

$$\int_\Omega |\nabla u|^2 = \int_\Omega |\nabla\psi|^2 + \int_\Omega |\nabla\Phi_2|^2$$

since ψ is constant on each component of $\partial\Omega$. Finally, we see after an expansion and integration by parts that

$$\int_\Omega |\nabla\psi|^2 \le 4\pi^2 \sum_{i=1}^n \operatorname{cap}\omega_i.$$

Using exactly the same argument as in the proof of Theorem I.3 we may handle still another useful variant. Assume, as above, that g_i, $i = 0, 1, 2, \ldots, n$, are given with

$$\deg(g_0, \partial G) = \sum_{i=1}^n \deg(g_i, \partial\omega_i);$$

and consider the class

$$\mathcal{E}'_2 = \left\{ v \in H^1(\Omega; S^1) \,\middle|\, \begin{array}{l} \forall i = 0, 1, \ldots, n, \ \exists \alpha_i \in \mathbb{C} \text{ with} \\ |\alpha_i| = 1 \text{ such that } v = \alpha_i g_i \text{ on } \partial\omega_i \end{array} \right\}$$

and set

(21) $$E'_2 = \operatorname*{Inf}_{v \in \mathcal{E}'_2} \int_\Omega |\nabla v|^2.$$

Theorem I.4. *The infimum in (21) is achieved by a map that is unique up to a phase. Moreover*

$$E_2' = \int_\Omega |\nabla \Phi_2|^2$$

where Φ_2 is the solution of (20).

I.3. S^1-valued harmonic maps with prescribed isolated singularities. The canonical harmonic map

As previously let G be a smooth, bounded and simply connected domain in $\mathbb{R}^2$. Fix n points $a_1, a_2, \dots, a_n$ in G, and n integers $d_1, d_2, \dots, d_n$ in $\mathbb{Z}$. Set $d = \sum_{i=1}^n d_i$ and let g be a map from ∂G into S^1, such that $\deg(g, \partial G) = d$. We propose to describe the class $\mathcal{C}$ of all smooth harmonic maps u from $\Omega = G \setminus \{a_1, \dots, a_n\}$ into S^1, such that:

(i) $\deg(u, a_i) = d_i$ (where $\deg(u, a_i)$ denotes the degree of u restricted to any small circle centered at a_i),

(ii) u is continuous up to ∂G and $u = g$ on ∂G.

Let Φ_0 be the solution of

$$\text{(22)} \qquad \begin{cases} \Delta \Phi_0 = \sum\limits_{i=1}^{n} 2\pi\, d_i \delta_{a_i} & \text{in } G, \\ \dfrac{\partial \Phi_0}{\partial \nu} = g \times g_\tau & \text{on } \partial G. \end{cases}$$

Since Φ_0 is unique up to an additive constant, we shall normalize Φ_0 by adding the condition that

$$\int_{\partial G} \Phi_0 = 0.$$

There exists a unique harmonic map u_0 in $\mathcal{C}$ associated to Φ_0, namely

$$\text{(23)} \qquad \begin{cases} u_0 \times \dfrac{\partial u_0}{\partial x_1} = -\dfrac{\partial \Phi_0}{\partial x_2} & \text{in } \Omega, \\ u_0 \times \dfrac{\partial u_0}{\partial x_2} = \dfrac{\partial \Phi_0}{\partial x_1} & \text{in } \Omega. \end{cases}$$

This follows from the fact that $\Delta \Phi_0 = 0$ in Ω, $\dfrac{\partial \Phi_0}{\partial \nu} = g \times g_\tau$ on ∂G, and $d_i \in \mathbb{Z}$ (see the arguments in the proof of Theorem I.2).

This u_0 will play an essential role. We shall call it the **canonical harmonic map** associated to (g, a, d).

Consider the class $\mathcal{L}$ of all real-valued harmonic functions ψ on Ω, that are smooth up to ∂G, possibly very singular at the points a_i, and equal to 0 on ∂G, i.e., $\psi \in \mathcal{L}$ means

$$\begin{cases} \Delta\psi = 0 & \text{in } \Omega, \\ \psi = 0 & \text{on } \partial G. \end{cases} \tag{24}$$

Theorem I.5. *There is a one-to-one correspondence between C and $\mathcal{L}$. More precisely, given any u in C, there is a unique ψ in $\mathcal{L}$ such that*

$$u = e^{i\psi}u_0. \tag{25}$$

Proof. Given u in C, consider the system

$$\begin{cases} \dfrac{\partial\hat{\psi}}{\partial x_1} = u \times \dfrac{\partial u}{\partial x_1} + \dfrac{\partial\Phi_0}{\partial x_2} = F_1 & \text{in } \Omega, \\ \dfrac{\partial\hat{\psi}}{\partial x_2} = u \times \dfrac{\partial u}{\partial x_2} - \dfrac{\partial\Phi_0}{\partial x_1} = F_2 & \text{in } \Omega. \end{cases} \tag{26}$$

The local existence of $\hat{\psi}$ follows from the fact that

$$\frac{\partial}{\partial x_2}\left(u \times \frac{\partial u}{\partial x_1} + \frac{\partial\Phi_0}{\partial x_2}\right) = \frac{\partial}{\partial x_1}\left(u \times \frac{\partial u}{\partial x_2} - \frac{\partial\Phi_0}{\partial x_1}\right).$$

In fact, $\hat{\psi}$ is globally defined, since for every closed curve C in Ω enclosing a domain ω,

$$\int_C F \cdot \tau = \int_C u \times u_\tau - \int_C \frac{\partial\Phi_0}{\partial\nu}$$

where ν is the outward unit normal to ω, and (ν, τ) is direct.

We have

$$\int_C u \times u_\tau = 2\pi \deg(u, C) = 2\pi \sum_{a_i \in \omega} d_i.$$

On the other hand (by (22)) we see that

$$\int_C \frac{\partial\Phi_0}{\partial\nu} = 2\pi \sum_{a_i \in \omega} d_i,$$

and therefore

$$\int_C F \cdot \tau = 0.$$

Finally, on ∂G, we have by (26) and (22)

$$\frac{\partial \hat{\psi}}{\partial \tau} = g \times g_\tau - \frac{\partial \Phi_0}{\partial \nu} = 0.$$

Thus $\hat{\psi}$ is constant on ∂G.

Set

$$\psi = \hat{\psi} - \hat{\psi}_{|\partial G}.$$

This ψ belongs to $\mathcal{L}$. Indeed, by (26),

$$\Delta \psi = \frac{\partial}{\partial x_1}\left(u \times \frac{\partial u}{\partial x_1}\right) + \frac{\partial}{\partial x_2}\left(u \times \frac{\partial u}{\partial x_2}\right) = u \times \Delta u = 0.$$

We claim that (25) holds. Indeed, we may write locally $u = e^{i\varphi}$ and $u_0 = e^{i\varphi_0}$.

Then (23) becomes

$$\frac{\partial \varphi_0}{\partial x_1} = -\frac{\partial \Phi_0}{\partial x_2},$$
$$\frac{\partial \varphi_0}{\partial x_2} = \frac{\partial \Phi_0}{\partial x_1}$$

and (26) becomes

$$\frac{\partial \psi}{\partial x_1} = \frac{\partial \varphi}{\partial x_1} + \frac{\partial \Phi_0}{\partial x_2},$$
$$\frac{\partial \psi}{\partial x_2} = \frac{\partial \varphi}{\partial x_2} - \frac{\partial \Phi_0}{\partial x_1}.$$

Thus $\nabla(\varphi_0 + \psi - \varphi) = 0$. Hence $\varphi_0 + \psi - \varphi$ is locally constant and vanishes on ∂G. It follows that $\varphi_0 + \psi - \varphi \equiv 0$. This completes the proof of the theorem.

Remark I.1. Since $|\nabla u_0| = |\nabla \Phi_0|$ and $|\nabla \Phi_0| \simeq \dfrac{|d_i|}{|x - a_i|}$ near a_i then $u_0 \in W^{1,1}(G)$ and even $u_0 \in W^{1,p}(G)$ for any $p < 2$, but $u_0 \notin W^{1,2}(G)$. Moreover u_0 satisfies the equation

$$\frac{\partial}{\partial x_1}\left(u_0 \times \frac{\partial u_0}{\partial x_1}\right) + \frac{\partial}{\partial x_2}\left(u_0 \times \frac{\partial u_0}{\partial x_2}\right) = 0 \quad \text{in } \mathcal{D}'(G). \tag{27}$$

More generally, let u be in $W^{1,1}(G) \cap \mathcal{C}$. The function ψ, which is associated to this u via Theorem I.5, belongs to $W^{1,1}(G)$ and we have

$$\frac{\partial}{\partial x_1}\left(u \times \frac{\partial u}{\partial x_1}\right) + \frac{\partial}{\partial x_2}\left(u \times \frac{\partial u}{\partial x_2}\right) = \Delta \psi \quad \text{in } \mathcal{D}'(G). \tag{28}$$

Since $\Delta\psi \in \mathcal{D}'(G)$, and $\text{supp}(\Delta\psi) \subset \bigcup_i \{a_i\}$, we know, by a celebrated result about distributions (see L. Schwartz [1]), that

$$\Delta\psi = \sum_{\text{finite}} c_{\alpha,i} D^\alpha(\delta_{a_i}).$$

In fact, here, we have only $\Delta\psi = 2\pi \sum c_i \delta_{a_i}$, because the left-hand side of (28) is a sum of (first order) derivatives of L^1 functions. Hence

$$\psi = \sum c_i \log|x - a_i| + \chi,$$

where χ is a smooth harmonic function on G, and the c_i's are some real constants. This means that any $u \in W^{1,1}(G) \cap C$ is of the form

$$u = u_0 e^{i\Sigma c_j \log|x-a_j|} e^{i\chi}.$$

In particular, u_0 is the unique element in $W^{1,1}(G) \cap C$ satisfying in addition (27).

Remark I.2. Connection with holomorphic and meromorphic functions on G. Given any u in C, there exists a (possibly multivalued) meromorphic function F on G, which is unique up to a positive multiplicative constant, such that

$$u = \frac{F}{|F|} \quad \text{in } \Omega. \tag{29}$$

Moreover, the zeroes and the singularities of F are necessarily contained in $\{a_1, \dots, a_n\}$. This follows from a result of Carbou [1]. Here is a simple proof. We may write locally $u = e^{i\varphi}$ where φ is harmonic. Then φ satisfies the following system:

$$\begin{cases} \dfrac{\partial\varphi}{\partial x_1} = -\dfrac{\partial\chi}{\partial x_2} \\ \dfrac{\partial\varphi}{\partial x_2} = \dfrac{\partial\chi}{\partial x_1}, \end{cases} \tag{30}$$

where χ is the local harmonic conjugate of φ. Thus $\chi + i\varphi$ is holomorphic. Hence

$$F = e^{\chi + i\varphi}$$

is also holomorphic, and

$$\frac{F}{|F|} = e^{i\varphi} = u.$$

A similar global construction yields a possibly multivalued function on G. Note that in the case where $u = u_0$, then F is single valued, and

$$F = \zeta(z - a_1)^{d_1} \dots (z - a_n)^{d_n} \tag{31}$$

where ζ is a nonvanishing holomorphic function on G. Indeed, in this case we may choose $\chi = \Phi_0$ in view of (23), which is uniquely globally defined on G, and $F = e^{\Phi_0} u_0$. We now study F near a point a_k, say $a_k = 0$.

By (22) we have

$$\Phi_0(x) = d_k \log |x| + R(x) \tag{32}$$

where R is a smooth harmonic function in some neighborhood of 0, including 0. From (23) we derive that

$$\begin{cases} u_0 \times \dfrac{\partial u_0}{\partial x_1} = -d_k \dfrac{x_2}{|x|^2} - \dfrac{\partial R}{\partial x_2} \\ u_0 \times \dfrac{\partial u_0}{\partial x_2} = \ \ d_k \dfrac{x_1}{|x|^2} + \dfrac{\partial R}{\partial x_1}. \end{cases} \tag{33}$$

Thus, we have, using polar coordinates,

$$u_0 = e^{id_k\theta} e^{iS} \tag{34}$$

where S is a harmonic function. Combining (32) and (34) we obtain

$$F = e^{\Phi_0} u_0 = z^{d_k} e^{R+iS}.$$

This proves (31).

Remark I.3. In view of the above discussion it is convenient to introduce the following definition:

Definition. Let G be an open set and let $a_1, a_2, \dots, a_n \in G$ be n points in G. Let $u : G \backslash \bigcup_i \{a_i\} \to S^1$ be a smooth harmonic map. Set $d_i = \deg(u, a_i)$. We say that u is **proper** if, for every i,

$$\lim_{z \to a_i} \frac{|z - a_i|^{d_i}}{(z - a_i)^{d_i}} u(z) \quad \text{exists.}$$

Note that, by (31), u_0 is a proper harmonic map.

Corollary I.1. *Assume G is a smooth, bounded, simply connected domain. Fix n points $a_1, a_2, \dots, a_n$ in G, fix n integers $d_1, d_2, \dots, d_n$ in $\mathbb{Z}$ and fix a boundary condition $g : \partial G \to S^1$ such that*

$$\deg(g, \partial G) = \sum_{i=1}^{n} d_i.$$

Then there exists a unique proper harmonic map such that

$$\deg(u, a_i) = d_i \quad \forall i \qquad \text{and } u = g \text{ on } \partial G.$$

Proof. The existence is clear (take $u = u_0$). We only need to prove uniqueness. By Theorem I.5 we know that any solution can be written as

$$u = e^{i\psi} u_0$$

for some real valued function ψ such that

$$\Delta\psi = 0 \quad \text{in } \Omega = G \setminus \bigcup_i \{a_i\}$$

and

$$\psi = 0 \quad \text{on } \partial G.$$

The condition that u is proper (since u_0 is proper) says that

$$\lim_{z \to a_i} e^{i\psi(z)} \text{ exists.}$$

This implies that $\lim_{z \to a_i} \psi(z)$ exists. By standard results it follows that $\psi \equiv 0$.

Finally, it is convenient to bear in mind the following construction of the canonical map u_0. Assume G, (a_i), (d_i) and g are as above.

Define, on ∂G,

$$\tilde{g}(z) = g(z) \frac{|z - a_1|^{d_1}}{(z - a_1)^{d_1}} \frac{|z - a_2|^{d_2}}{(z - a_2)^{d_2}} \cdots \frac{|z - a_n|^{d_n}}{(z - a_n)^{d_n}}.$$

Note that $\deg(\tilde{g}, \partial G) = 0$ and hence there is a well defined single-valued (smooth) function $\varphi_0 : \partial G \to \mathbb{R}$ such that

$$\tilde{g} = e^{i\varphi_0} \quad \text{on } \partial G.$$

We also denote by φ_0 its harmonic extension in G.

Corollary I.2. *We have, on G,*

$$u_0(z) = e^{i\varphi_0(z)} \frac{(z-a_1)^{d_1}}{|z-a_1|^{d_1}} \frac{(z-a_2)^{d_2}}{|z-a_2|^{d_2}} \cdots \frac{(z-a_n)^{d_n}}{|z-a_n|^{d_n}}.$$

Proof. Note that the right-hand side is a proper harmonic map satisfying the appropriate boundary condition on ∂G and degree conditions at the a_i's. By Corollary I.1 it must coincide with u_0.

Remark I.4. It is sometimes more convenient to work with the expression for u_0 given in Corollary I.2 rather than the construction of u_0 via Φ_0 (see (23)). In Corollary I.2 u_0 is related to the solution of a **Dirichlet** linear problem while in (23) u_0 is related (via Φ_0) to the solution of a **Neumann** linear problem. In some concrete situations where an explicit representation formula for u_0 is required, it is easier to work with the Dirichlet condition (see e.g., Section VIII.4).

I.4. Shrinking holes. Renormalized energy

As in Section I.3, let G be a smooth, bounded and simply connected domain in $\mathbb{R}^2$. Fix n points $a_1, \ldots, a_n$ in G, and n integers $d_1, \ldots, d_n$ in $\mathbb{Z}$. Set $d = \sum_{i=1}^{n} d_i$, and let g be a map from ∂G to S^1 such that $\deg(g, \partial G) = d$. Let ρ be a positive constant that we will let tend to 0. Set $\Omega_\rho = G \setminus \bigcup_{i=1}^{n} \overline{B(a_i, \rho)}$, and consider the minimization problem

$$\operatorname*{Min}_{u \in \mathcal{E}_\rho} \int_{\Omega_\rho} |\nabla u|^2,$$

where $\mathcal{E}_\rho = \{v \in H^1(\Omega_\rho; S^1);\ \deg(v, \partial B(a_i, \rho)) = d_i \text{ and } v = g \text{ on } \partial G\}$. We already know (by Theorem I.2) that there exists a unique minimizer, which we denote u_ρ.

Theorem I.6. *As ρ tends to 0, u_ρ converges to u_0 (the canonical harmonic map) uniformly on every compact subset of $\Omega \bigcup \partial G$.*

Proof. Let Φ_0 be the solution of (22) and Φ_ρ be the solution of

$$\begin{cases} \Delta \Phi_\rho = 0 & \text{in } \Omega_\rho, \\ \Phi_\rho = \text{Const.} & \text{on each } \partial\omega_i \text{ with } \omega_i = B(a_i, \rho), \\ \displaystyle\int_{\partial \omega_i} \frac{\partial \Phi_\rho}{\partial \nu} = 2\pi\, d_i & i = 1, 2, \ldots, n, \\ \displaystyle\frac{\partial \Phi_\rho}{\partial \nu} = g \times g_\tau & \text{on } \partial G. \\ \displaystyle\int_{\partial G} \Phi_\rho = 0. \end{cases}$$

Lemma I.2. *As ρ tends to zero, Φ_ρ converges to Φ_0, uniformly. More precisely, we have*

$$\|\Phi_\rho - \Phi_0\|_{L^\infty(\Omega_\rho)} \le C\rho.$$

The proof is based on the following lemma.

Lemma I.3. *Let G be a smooth bounded domain in $\mathbb{R}^2$, and let ω_i, $i = 1, 2, \ldots, n$, be smooth disjoint subdomains of G, such that $\Omega = G \setminus \bigcup_{i=1}^{n} \bar{\omega}_i$ is connected. Let v be a function satisfying*

$$\begin{cases} \Delta v = 0 & \text{in } \Omega, \\ \displaystyle\int_{\partial\omega_i} \frac{\partial v}{\partial \nu} = 0 & \text{for each } i = 1, 2, \ldots, n, \\ \displaystyle\frac{\partial v}{\partial \nu} = 0 & \text{on } \partial G. \end{cases}$$

Then

$$\operatorname*{Sup}_{\Omega} v - \operatorname*{Inf}_{\Omega} v \le \sum_{i=1}^{n} (\operatorname*{Sup}_{\partial\omega_i} v - \operatorname*{Inf}_{\partial\omega_i} v). \tag{35}$$

Proof of Lemma I.3. Set

$$I_i = [\alpha_i, \beta_i] \quad \text{where } \alpha_i = \operatorname*{Inf}_{\partial\omega_i} v \text{ and } \beta_i = \operatorname*{Sup}_{\partial\omega_i} v.$$

We claim that

$$\bigcup_{i=1}^{n} I_i \text{ is connected.} \tag{36}$$

For otherwise there would be some $t_0 \in \mathbb{R}$ and some integer k, $1 \le k < n$, such that (after relabelling the intervals)

$$\beta_i \le t_0 - \delta \qquad i = 1, 2, \ldots, k$$

and

$$\alpha_i \ge t_0 + \delta \qquad i = k+1, \ldots, n,$$

for some $\delta > 0$. Choose any smooth function $\theta : \mathbb{R} \to \mathbb{R}$ such that

$$\theta(t) = \begin{cases} 0 & \text{if } t \le t_0 - \delta \\ 1 & \text{if } t \ge t_0 + \delta \end{cases}$$

and $\theta'(t) > 0\ \forall t \in (t_0 - \delta, t_0 + \delta)$. We have

$$0 = -\int_\Omega (\Delta v)\theta(v) = \int_\Omega \theta'(v)|\nabla v|^2 - \int_{\partial G} \frac{\partial v}{\partial \nu}\,\theta(v) + \sum_{i=1}^n \int_{\partial \omega_i} \frac{\partial v}{\partial \nu}\,\theta(v).$$

But $\dfrac{\partial v}{\partial \nu} = 0$ on ∂G, $\theta(v)$ is constant (either 0 or 1) on each $\partial\omega_i$ and

$$\int_{\partial\omega_i} \frac{\partial v}{\partial \nu} = 0 \quad \text{for } i = 1, 2, \dots, n.$$

Thus we obtain

$$\int_\Omega \theta'(v)|\nabla v|^2 = 0$$

and therefore $\nabla v = 0$ on the set $B = \{x \in \Omega; t_0 - \delta < v(x) < t_0 + \delta\}$. Hence v is locally constant on B and consequently it assumes at most a countable number of values between $t_0 - \delta$ and $t_0 + \delta$. On the other hand, Ω is connected and thus v assumes all its values between $\operatorname*{Inf}_\Omega v$ and $\operatorname*{Sup}_\Omega v$. In particular, v assumes all its values between $t_0 - \delta$ and $t_0 + \delta$. This yields a contradiction and the proof of (36) is complete.

We now turn to the proof of (35). From (36) we deduce that

$$\operatorname*{Max}_{1 \le i \le n} \beta_i - \operatorname*{Min}_{1 \le i \le n} \alpha_i \le \sum_{i=1}^n (\beta_i - \alpha_i). \tag{37}$$

By adding a constant to v we may always assume that $\operatorname*{Min}_{1 \le i \le n} \alpha_i = 0$. Set

$$A = \operatorname*{Max}_{1 \le i \le n} \beta_i.$$

Multiplying the equation $\Delta v = 0$ by $(v - A)^+$ and integrating we see that

$$\int_\Omega |\nabla (v - A)^+|^2 + \sum_{i=1}^n \int_{\partial\omega_i} \frac{\partial v}{\partial \nu}(v - A)^+ = 0.$$

But $(v - A)^+ = 0$ on each $\partial\omega_i$ and therefore $(v - A)^+ \equiv 0$ on Ω, i.e., $v \le A$ on Ω. Similarly $v^- = 0$ on each $\partial\omega_i$ and therefore $v \ge 0$ on Ω. Hence we obtain

$$\operatorname*{Inf}_\Omega v = 0$$

and

$$\operatorname*{Sup}_\Omega v \le A = \operatorname*{Max}_{1 \le i \le n} \beta_i \le \sum_{i=1}^n (\beta_i - \alpha_i) \text{ by (37).}$$

This completes the proof of (35).

A variant of Lemma I.3 that we find useful is the following:

Lemma I.4. *Let G be a smooth, bounded domain in $\mathbb{R}^2$ and let (ω_i) be smooth disjoint subdomains of G such that $\Omega = G\setminus\bigcup_{i=1}^{n}\bar{\omega}_i$ is connected. Let v be a function satisfying*

$$\left\{\begin{array}{ll}\Delta v = 0 & \text{in } \Omega\\ \displaystyle\int_{\partial\omega_j}\frac{\partial v}{\partial \nu} = 0 & \text{for } j = 1,2,\ldots,n.\end{array}\right. \tag{38}$$

Then

$$\operatorname*{Sup}_{\Omega} v - \operatorname*{Inf}_{\Omega} v \le \sum_{j=1}^{n}\left(\operatorname*{Sup}_{\partial\omega_j} v - \operatorname*{Inf}_{\partial\omega_j} v\right) + \operatorname*{Sup}_{\partial G} v - \operatorname*{Inf}_{\partial G} v. \tag{39}$$

In particular if $v = 0$ on ∂G then

$$\|v\|_{L^\infty(\Omega)} \le \sum_{j=1}^{n}\left(\operatorname*{Sup}_{\partial\omega_j} v - \operatorname*{Inf}_{\partial\omega_j} v\right). \tag{40}$$

Note that, here, we do not assume (as in Lemma I.3) that $\dfrac{\partial v}{\partial \nu} = 0$ on ∂G; but, on the other hand, the conclusion is weaker and also involves the oscillation of v on ∂G. The proof of Lemma I.4 is almost identical to the proof of Lemma I.3 and we shall omit it. It is convenient to introduce $I_0 = [\alpha_0, \beta_0]$ where $\alpha_0 = \operatorname*{Inf}_{\partial G} v$ and $\beta_0 = \operatorname*{Sup}_{\partial G} v$ and to follow the same argument as in the proof of Lemma I.3 where the index i runs between 0 and n.

We may now return to:

Proof of Lemma I.2. We apply Lemma I.3 to $v = \Phi_\rho - \Phi_0$ on Ω_ρ. Since $\Phi_\rho = \text{Const.}$ on each $\partial B(a_i, \rho)$ we have

$$\begin{aligned}&\operatorname*{Sup}_{\Omega_\rho}(\Phi_\rho - \Phi_0) - \operatorname*{Inf}_{\Omega_\rho}(\Phi_\rho - \Phi_0)\\ &\qquad \le \sum_{i=1}^{n} \operatorname*{Sup}_{\partial B(a_i,\rho)} \Phi_0 - \operatorname*{Inf}_{\partial B(a_i,\rho)} \Phi_0 \le C\rho.\end{aligned} \tag{41}$$

On the other hand, we know that

$$\int_{\partial G}(\Phi_\rho - \Phi_0) = 0,$$

and thus there is a point on ∂G where $\Phi_\rho - \Phi_0 = 0$. From this and (41), we deduce that

$$\|\Phi_\rho - \Phi_0\|_{L^\infty(\Omega_\rho)} \le C\rho. \tag{42}$$

Proof of Theorem I.6 completed. It follows from Lemma I.2 and elliptic estimates that

$$\Phi_\rho \to \Phi_0 \quad \text{in } C^k_{\text{loc}}(\Omega \cup \partial G)$$

and

$$\|\Phi_\rho - \Phi_0\|_{C^k(K)} \le C\rho \tag{43}$$

for any compact set K of $\Omega \bigcup \partial G$. We recall that

$$\begin{aligned} u_\rho \times (u_\rho)_{x_1} &= -\frac{\partial \Phi_\rho}{\partial x_2} && \text{in } \Omega_\rho \\ u_\rho \times (u_\rho)_{x_2} &= \frac{\partial \Phi_\rho}{\partial x_1} && \text{in } \Omega_\rho \end{aligned}$$

and

$$\begin{aligned} u_0 \times (u_0)_{x_1} &= -\frac{\partial \Phi_0}{\partial x_2} && \text{in } \Omega \\ u_0 \times (u_0)_{x_2} &= \frac{\partial \Phi_0}{\partial x_1} && \text{in } \Omega. \end{aligned}$$

Clearly (43) implies that

$$\|u_\rho - u_0\|_{C^k(K)} \le C\rho. \tag{44}$$

This completes the proof of Theorem I.6.

Theorem I.7. *Set*

$$R_0(x) = \Phi_0(x) - \sum_{j=1}^{n} d_j \log|x - a_j| \tag{45}$$

so that R_0 is a smooth harmonic function on G. Then, as $\rho \to 0$,

$$\frac{1}{2}\int_{\Omega_\rho} |\nabla u_\rho|^2 = \pi \left(\sum_{i=1}^{n} d_i^2\right) \log(1/\rho) + W + O(\rho) \tag{46}$$

where

$$W = -\pi \sum_{i \neq j} d_i d_j \log |a_i - a_j| + \frac{1}{2} \int_{\partial G} \Phi_0 (g \times g_\tau) - \pi \sum_{i=1}^{n} d_i R_0(a_i)$$

(47)

$$= W(a, d, g).$$

Note that W is independent of ρ and depends only on G, (a_i), (d_i) and g. W will be called the **"renormalized energy"**. Here $O(\rho)$ stands for a quantity X such that $|X| \leq C\rho$ where C depends only on G, (a_i), (d_i) and g.

Proof. By Theorem I.2 we know that

$$\int_{\Omega_\rho} |\nabla u_\rho|^2 = \int_{\Omega_\rho} |\nabla \Phi_\rho|^2.$$

Since Φ_ρ is harmonic in Ω_ρ we have

$$\int_{\Omega_\rho} |\nabla \Phi_\rho|^2 = \int_{\partial G} \frac{\partial \Phi_\rho}{\partial \nu} \Phi_\rho - \sum_{i=1}^{n} \int_{\partial B(a_i, \rho)} \frac{\partial \Phi_\rho}{\partial \nu} \Phi_\rho.$$

Recall that $\dfrac{\partial \Phi_\rho}{\partial \nu} = g \times g_\tau$ on $\partial G, \Phi_\rho = \text{Const.}$ on $\partial B(a_i, \rho)$ and that

$$\int_{\partial B(a_i, \rho)} \frac{\partial \Phi_\rho}{\partial \nu} = 2\pi \, d_i.$$

Thus, we obtain

$$\int_{\Omega_\rho} |\nabla \Phi_\rho|^2 = \int_{\partial G} \Phi_\rho (g \times g_\tau) - \sum_{i=1}^{n} 2\pi \, d_i \Phi_\rho(\partial B(a_i, \rho)).$$

From (42), we deduce that

$$\int_{\Omega_\rho} |\nabla \Phi_\rho|^2 = \int_{\partial G} \Phi_0 (g \times g_\tau) - \sum_{i=1}^{n} 2\pi \, d_i \Phi_0(x_i) + O(\rho) \tag{48}$$

where x_i is any point on $\partial B(a_i, \rho)$. Inserting (45) in (48), we are led to (46).

Since the renormalized energy W plays an important role in our analysis it will be convenient to have several other expressions for W.

Theorem I.8. *Let u_0 be the canonical harmonic map associated to (g, a, d), then, as $\rho \to 0$,*

$$\frac{1}{2}\int_{\Omega_\rho} |\nabla u_0|^2 = \pi \left(\sum_{i=1}^{n} d_i^2\right) \log(1/\rho) + W + O(\rho^2) \tag{49}$$

where $\Omega_\rho = G \setminus \bigcup_{i=1}^{n} \overline{B(a_i, \rho)}$.

Proof. Note that, by (23), $|\nabla u_0| = |\nabla \Phi_0|$ and therefore,

$$\int_{\Omega_\rho} |\nabla u_0|^2 = \int_{\Omega_\rho} |\nabla \Phi_0|^2 = \int_{\partial G} \frac{\partial \Phi_0}{\partial \nu}\Phi_0 - \sum_{i=1}^{n} \int_{\partial B(a_i,\rho)} \frac{\partial \Phi_0}{\partial \nu}\Phi_0. \tag{50}$$

Recall that, (see (22)),

$$\frac{\partial \Phi_0}{\partial \nu} = g \times g_\tau \quad \text{on } \partial G$$

and that

$$R_0(x) = \Phi_0(x) - \sum_{j=1}^{n} d_j \log|x - a_j|$$

is a smooth harmonic function on G.

Set

$$S_i(x) = \Phi_0(x) - d_i \log|x - a_i|. \tag{51}$$

Note that S_i is a smooth harmonic function in some neighborhood of a_i (including a_i) and that

$$S_i(x) = \Phi_0(x) - d_i \log \rho \quad \text{on } \partial B(a_i, \rho),$$

$$\frac{\partial S_i}{\partial \nu} = \frac{\partial \Phi_0}{\partial \nu} - \frac{d_i}{\rho} \quad \text{on } \partial B(a_i, \rho)$$

and

$$S_i(a_i) = R_0(a_i) + \sum_{j \neq i} d_j \log|a_i - a_j|.$$

Thus, we have

$$\int_{\partial B(a_i,\rho)} \frac{\partial \Phi_0}{\partial \nu}\Phi_0 = \int_{\partial B(a_i,\rho)} \left(\frac{\partial S_i}{\partial \nu} + \frac{d_i}{\rho}\right)(S_i + d_i \log \rho)$$

and, using the fact that S_i is harmonic in $B(a_i,\rho)$, we deduce that

$$\int_{\partial B(a_i,\rho)} \frac{\partial \Phi_0}{\partial \nu}\Phi_0 = \int_{B(a_i,\rho)} |\nabla S_i|^2 + 2\pi\, d_i S_i(a_i) + 2\pi\, d_i^2 \log\,\rho. \tag{52}$$

Combining (50) and (52) we are led to

$$\int_{\Omega_\rho} |\nabla u_0|^2 = 2\pi\left(\sum_{i=1}^{n} d_i^2\right)\log(1/\rho) + 2W - \sum_{i=1}^{n}\int_{B(a_i,\rho)} |\nabla S_i|^2 \tag{53}$$

which yields the desired conclusion.

Next, we present a variant of Theorem I.7 which will be useful in Section VIII.1. Consider the minimization problem

$$\operatorname*{Min}_{u\in\hat{\mathcal{E}}_\rho} \int_{\Omega_\rho} |\nabla u|^2$$

where

$$\hat{\mathcal{E}}_\rho = \left\{ v \in H^1(\Omega_\rho; S^1) \,\middle|\, \begin{array}{l} v = g \text{ on } \partial G \text{ and } \forall i,\ \exists \alpha_i \in \mathbb{C} \text{ with } |\alpha_i| = 1 \\ \text{such that } v(z) = \dfrac{\alpha_i}{\rho^{d_i}}(z-a_i)^{d_i} \text{ on } \partial B(a_i,\rho) \end{array} \right\}.$$

We already know, by Theorem I.4, that there is a unique minimizer, which we denote $\hat{u}_\rho$.

Theorem I.9. *We have, as* $\rho \to 0$,

$$\frac{1}{2}\int_{\Omega_\rho} |\nabla \hat{u}_\rho|^2 = \pi\left(\sum_{i=1}^{n} d_i^2\right)\log(1/\rho) + W + O(\rho). \tag{54}$$

Proof. By Theorem I.4 we know that

$$\int_{\Omega_\rho} |\nabla \hat{u}_\rho|^2 = \int_{\Omega_\rho} |\nabla \hat{\Phi}_\rho|^2 \tag{55}$$

where $\hat{\Phi}_\rho$ is the solution of the linear problem

$$\begin{cases} \Delta\hat{\Phi}_\rho = 0 & \text{in } \Omega_\rho, \\ \dfrac{\partial \hat{\Phi}_\rho}{\partial \nu} = g \times g_\tau & \text{on } \partial G, \\ \dfrac{\partial \hat{\Phi}_\rho}{\partial \nu} = \dfrac{d_i}{\rho} & \text{on } \partial B(a_i,\rho),\ i = 1,2,\ldots,n. \end{cases} \tag{56}$$

Since $\hat{\Phi}_\rho$ is unique up to an additive constant we may normalize it by assuming that

$$\int_{\partial G} \hat{\Phi}_\rho = 0.$$

Set

$$\Psi_\rho(x) = \hat{\Phi}_\rho(x) - \sum_{j=1}^{n} d_j \log|x - a_j|. \tag{57}$$

We shall use the following lemma, the proof of which was suggested to us by L. Nirenberg:

Lemma I.5. *We have, as $\rho \to 0$,*

$$\|\Psi_\rho - R_0\|_{L^\infty(\partial G)} \le C\rho \tag{58}$$

and

$$\|\Psi_\rho\|_{L^\infty(\partial B(a_i,\rho))} \le C. \tag{59}$$

Proof. The function Ψ_ρ satisfies

$$\begin{cases} \Delta\Psi_\rho = 0 & \text{in } \Omega_\rho, \\ \dfrac{\partial \Psi_\rho}{\partial \nu} = g \times g_\tau - \sum_{j=1}^{n} d_j \dfrac{\partial}{\partial \nu} \log|x - a_j| \equiv f & \text{on } \partial G, \\ \dfrac{\partial \Psi_\rho}{\partial \nu} = -\sum_{j \ne i} d_j \dfrac{\partial}{\partial \nu} \log|x - a_j| \equiv g_i, \forall i & \text{on } \partial B(a_i, \rho). \end{cases} \tag{60}$$

Let $\Psi_\rho^\star$ be a harmonic conjugate of Ψ_ρ, i.e., $\Psi_\rho^\star$ is a solution of

$$\begin{cases} \dfrac{\partial \Psi_\rho^\star}{\partial x_1} = -\dfrac{\partial \Psi_\rho}{\partial x_2} & \text{in } \Omega_\rho, \\ \dfrac{\partial \Psi_\rho^\star}{\partial x_2} = \dfrac{\partial \Psi_\rho}{\partial x_1} & \text{in } \Omega_\rho. \end{cases} \tag{61}$$

Note that $\Psi_\rho^\star$ is well defined globally (see Lemma I.1) on Ω_ρ as a single-valued function since

$$\int_\Gamma \frac{\partial \Psi_\rho}{\partial \nu} = 0$$

for each connected component Γ of $\partial\Omega_\rho$. The function $\Psi_\rho^\star$ satisfies

$$\begin{cases} \Delta\Psi_\rho^\star = 0 & \text{in } \Omega_\rho, \\ \dfrac{\partial \Psi_\rho^\star}{\partial \tau} = f & \text{on } \partial G, \\ \dfrac{\partial \Psi_\rho^\star}{\partial \tau} = g_i & \text{on } \partial B(a_i, \rho),\ i = 1, 2, \ldots, n. \end{cases}$$

Hence, we have

$$
\tag{62}
\begin{cases}
\Delta\Psi^\star_\rho = 0 & \text{in } \Omega_\rho,\\
\Psi^\star_\rho = F & \text{on } \partial G,\\
\Psi^\star_\rho = G_i & \text{on } \partial B(a_i,\rho),\ \ i=1,2,\dots,n,\\
\displaystyle\int_{\partial B(a_i,\rho)} \frac{\partial \Psi^\star_\rho}{\partial\nu} = 0 & i=1,2,\dots,n,
\end{cases}
$$

where F (resp. G_i) is a primitive with respect to arc length of f (resp. g_i) on ∂G (resp. $\partial B(a_i,\rho)$), i.e.,

$$\frac{\partial F}{\partial \tau} = f \qquad \left(\text{resp. } \frac{\partial G_i}{\partial \tau} = g_i\right).$$

[Note that F (resp. G_i) is well defined as a single-valued function since $\int_{\partial G} f = 0$ (resp. $\int_{\partial B(a_i,\rho)} g_i = 0$)].

Let $\Psi^\star$ be the solution of

$$
\tag{63}
\begin{cases}
\Delta\Psi^\star = 0 & \text{in } G,\\
\Psi^\star = F & \text{on } \partial G.
\end{cases}
$$

Applying Lemma I.4 to $v = \Psi^\star_\rho - \Psi^\star$ we see that

$$
\tag{64}
\|\Psi^\star_\rho - \Psi^\star\|_{L^\infty(\Omega_\rho)} \le \sum_{i=1}^n \operatorname*{Sup}_{\partial B(a_i,\rho)} (G_i - \Psi^\star) - \operatorname*{Inf}_{\partial B(a_i,\rho)} (G_i - \Psi^\star) = O(\rho)
$$

since $\|g_i\|_{L^\infty(\partial B(a_i,\rho))} \le C$.

From (62), (63), (64) and standard elliptic estimates (see e.g., D. Gilbarg and N. Trudinger [1]) we deduce that for every compact subset $K \subset \bar G \setminus \bigcup_i \{a_i\}$

$$
\tag{65}
\|\nabla(\Psi^\star_\rho - \Psi^\star)\|_{L^\infty(K)} \le C_K \rho.
$$

Let Ψ be a harmonic conjugate of $\Psi^\star$, i.e.,

$$
\tag{66}
\begin{cases}
\dfrac{\partial \Psi}{\partial x_1} = \dfrac{\partial \Psi^\star}{\partial x_2} & \text{in } G\\[2mm]
\dfrac{\partial \Psi}{\partial x_2} = -\dfrac{\partial \Psi^\star}{\partial x_1} & \text{in } G
\end{cases}
$$

so that Ψ satisfies

$$\begin{cases} \Delta\Psi = 0 & \text{in } G \\ \dfrac{\partial\Psi}{\partial\nu} = \dfrac{\partial\Psi^\star}{\partial\tau} = f & \text{on } \partial G. \end{cases} \tag{67}$$

Recall that R_0 also satisfies (67) and since the solution of (67) is unique up to an additive constant we may as well choose

$$\Psi = R_0. \tag{68}$$

From (61) and (66) we deduce that

$$|\nabla(\Psi^\star_\rho - \Psi^\star)| = |\nabla(\Psi_\rho - \Psi)|$$

and, using (65) together with (68), we conclude that

$$\|\nabla(\Psi_\rho - R_0)\|_{L^\infty(K)} \le C_K\rho. \tag{69}$$

Finally, we recall that, by our normalization choice,

$$\int_{\partial G} (\Psi_\rho - R_0) = 0 \tag{70}$$

(since $\int_{\partial G} \hat{\Phi}_\rho = \int_{\partial G} \Phi_0 = 0$).
It follows from (69) and (70) that

$$\|\Psi_\rho - R_0\|_{L^\infty(K)} \le C_K\rho. \tag{71}$$

In particular, we have proved (58).

We now turn to the proof of (59). Set

$$\omega_i(x) = \sum_{j\ne i} d_j \log|x - a_j|.$$

Fix any $\alpha > 0$ so that $\overline{B(a_i,\alpha)} \subset G$ and $\overline{B(a_i,\alpha)}$ does not contain any other point a_j, $j \ne i$. From (60) we deduce that

$$\Delta(\Psi_\rho + \omega_i) = 0 \quad \text{in } B(a_i,\alpha) \setminus B(a_i,\rho),$$

and

$$\frac{\partial}{\partial\nu}(\Psi_\rho + \omega_i) = 0 \quad \text{on } \partial B(a_i,\rho).$$

It follows from the maximum principle that

$$\|\Psi_\rho + \omega_i\|_{L^\infty(B(a_i,\alpha)\setminus B(a_i,\rho))} \le \|\Psi_\rho + \omega_i\|_{L^\infty(\partial B(a_i,\alpha))} \le C$$

by (71). In particular,

$$\|\Psi_\rho + \omega_i\|_{L^\infty(\partial B(a_i,\rho))} \le C$$

and this implies (59).

Proof of Theorem I.9 completed. We may now turn to (54). Using (55) and (57) we write

$$\int_{\Omega_\rho} |\nabla \hat{u}_\rho|^2 = \int_{\Omega_\rho} \left|\nabla\left(\Psi_\rho + \sum_{j=1}^n d_j \log|x-a_j|\right)\right|^2.$$

Therefore we have

$$(72) \quad \int_{\Omega_\rho} |\nabla \hat{u}_\rho|^2 = \int_{\Omega_\rho} |\nabla\Psi_\rho|^2 + 2\int_{\Omega_\rho} \nabla\Psi_\rho \nabla\left(\sum_{j=1}^n d_j \log|x-a_j|\right)$$
$$+ \int_{\Omega_\rho} \left|\nabla\left(\sum_{j=1}^n d_j \log|x-a_j|\right)\right|^2.$$

Integrating by parts, using (60) and Lemma I.5 we obtain

$$(73) \quad \int_{\Omega_\rho} |\nabla\Psi_\rho|^2 = \int_{\partial G} f\Psi_\rho - \sum_{i=1}^n \int_{\partial B(a_i,\rho)} g_i\Psi_\rho = \int_{\partial G} f\, R_0 + O(\rho)$$
$$= \int_G |\nabla R_0|^2 + O(\rho) = \int_{\Omega_\rho} |\nabla R_0|^2 + O(\rho)$$

and

$$(74) \quad \int_{\Omega_\rho} \nabla(\Psi_\rho - R_0)\nabla\left(\sum_{j=1}^n d_j \log|x-a_j|\right)$$
$$= -\sum_{i=1}^n \int_{\partial B(a_i,\rho)} \left(g_i - \frac{\partial R_0}{\partial \nu}\right)\omega_i = O(\rho).$$

Combining (72), (73) and (74) we are led to

$$\int_{\Omega_\rho} |\nabla \hat{u}_\rho|^2 = \int_{\Omega_\rho} \left|\nabla \left(R_0 + \sum_{j=1}^{n} d_j \log|x - a_j| \right)\right|^2 + O(\rho)$$
$$= \int_{\Omega_\rho} |\nabla \Phi_0|^2 + O(\rho) = \int_{\Omega_\rho} |\nabla u_0|^2 + O(\rho).$$

Applying Theorem I.8 we obtain the desired conclusion.

Remark I.5. One could also work with a still more restrictive class of testing maps:

We have, as $\rho \to 0$,

$$\operatorname*{Min}_{u \in \tilde{\mathcal{E}}_\rho} \frac{1}{2} \int_{\Omega_\rho} |\nabla u|^2 = \pi \left(\sum_{i=1}^{n} d_i^2 \right) \log(1/\rho) + W + O(1/|\log \rho|)$$

where

$$\tilde{\mathcal{E}}_\rho = \left\{ v \in H^1(\Omega_\rho; S^1) \,\middle|\, \begin{array}{l} v = g \quad \text{on } \partial G \quad \text{and} \\ v(z) = \dfrac{(z - a_i)^{d_i}}{\rho^{d_i}} \text{ on } \partial B(a_i, \rho),\ \forall i \end{array} \right\}.$$

This follows from Theorems I.3 and I.9 and the fact that $\operatorname{cap}(B(a_i, \rho)) = O(1/|\log \rho|)$.

Remark I.6. Suppose now that the integer n and the points (a_i) are not prescribed: they are free to move in G. Suppose that the degrees (d_i) are not given; they are only constrained by the relation $\sum_{i=1}^{n} d_i = d = \deg(g, \partial G)$ where $g : \partial G \to S^1$ is given. If we want to minimize $\int_{\Omega_\rho} |\nabla u_\rho|^2$ (for ρ small) among all possible choices of n, (a_i) and (d_i) we are led to:

(i) Choose $n = d$ and each $d_i = +1$. This follows from the obvious fact that

$$\operatorname{Min} \left\{ \sum_{i=1}^{n} d_i^2 \,;\, \sum_{i=1}^{n} d_i = d \right\}$$

is achieved when $n = d$ and each $d_i = +1$.

(ii) Choose a configuration (a_i) that minimizes W.

The existence of a minimizer for W is guaranteed by the following:

Theorem I.10. *Assume $d_i = +1\ \forall i$. Fix a boundary condition g and consider $W = W(a)$ only as a function of the configuration a.*

Then

$$W(a) \to +\infty \quad \text{as} \quad \min\left\{\min_{i\neq j}|a_i - a_j|, \min_i \operatorname{dist}(a_i, \partial G)\right\} \to 0.$$

In other words, $W \to +\infty$ as two of the points (a_i) coalesce or as one of the points a_i tends to ∂G. Therefore, Min W is achieved and every minimizing configuration consists of d distinct points in G^d (not $\bar{G}^d$).

Theorem I.10 is a consequence of the following two lemmas and the explicit expression for W given by (47).

Lemma I.6. *There is a constant C independent of a such that*

$$\int_{\partial G} |\Phi_0| \leq C.$$

Proof. Let ψ be the solution of

$$\begin{cases} \Delta\psi = 2\pi \sum_{i=1}^d \delta_{a_i} & \text{in } G \\ \psi = 0 & \text{on } \partial G \end{cases}$$

so that, by the maximum principle, $\psi \leq 0$ in G and $\dfrac{\partial\psi}{\partial\nu} \geq 0$ on ∂G. Thus,

$$\int_{\partial G} \left|\frac{\partial\psi}{\partial\nu}\right| = \int_{\partial G} \frac{\partial\psi}{\partial\nu} = 2\pi\, d.$$

On the other hand, we have

$$\begin{cases} \Delta(\Phi_0 - \psi) = 0 & \text{in } G, \\ \dfrac{\partial}{\partial\nu}(\Phi_0 - \psi) = (g \times g_\tau) - \dfrac{\partial\psi}{\partial\nu} & \text{on } \partial G. \end{cases}$$

It follows from standard elliptic estimates that

$$\|\Phi_0 - \psi\|_{L^1(\partial G)} \leq C\left\|(g \times g_\tau) - \frac{\partial\psi}{\partial\nu}\right\|_{L^1(\partial G)} \leq C$$

which yields the desired estimate for Φ_0.

Lemma I.7. *We have*

$$R_0(a) \to -\infty \quad \text{as } \min_i \operatorname{dist}(a_i, \partial G) \to 0.$$

Proof. For simplicity, we shall only consider where $d = 1$ (i.e., the configuration (a_i) consists of a single point). We shall sketch the proof in the case of a flat boundary. Assume that $0 \in \partial G$ and that, locally, near 0, G is the half-plane, $G = \{(x_1, x_2);\ x_1 < 0\}$. Let (a_n) be a sequence in G such that $a_n \to 0$. Consider the function

$$v_n(x) = \log|x - a_n| + \log|x - a_n^\star| + \alpha_n \tag{75}$$

where $a_n^\star$ is the reflected point of a_n about the x_2-axis and α_n is a constant chosen such that

$$\int_{\partial G} v_n = 0.$$

Note that $|\alpha_n| \le C$ since $\int_{\partial G} \big|\log|x|\big| < \infty$. We have (calling Φ_n the function Φ_0 associated to a_n),

$$\begin{cases} \Delta(\Phi_n - v_n) = 0 & \text{in } G, \\ \dfrac{\partial}{\partial\nu}(\Phi_n - v_n) = (g \times g_\tau) - \dfrac{\partial v_n}{\partial \nu} & \text{on } \partial G. \end{cases} \tag{76}$$

Using (75) we see that $\dfrac{\partial v_n}{\partial \nu} = 0$ on $\{x_1 = 0\} \cap \partial G$; it follows easily that

$$\left\| \frac{\partial v_n}{\partial \nu} \right\|_{L^\infty(\partial G)} \le C.$$

From (76) and standard elliptic estimates we conclude that

$$\|\Phi_n - v_n\|_{L^\infty(G)} \le C.$$

Recall that $R_n(x) = \Phi_n(x) - \log|x - a_n|$ and thus

$$\big\|R_n(x) - \log|x - a_n^\star|\big\|_{L^\infty(G)} \le C.$$

In particular, we have

$$\big|R_n(a_n) - \log|a_n - a_n^\star|\big| \le C$$

and the desired conclusion follows since $\log|a_n - a_n^\star| \to -\infty$.

CHAPTER II

A lower bound for the energy of S^1-valued maps on perforated domains

Let $G \subset \mathbb{R}^2$ be a smooth, bounded and connected domain. Let $x_1, x_2, \ldots, x_n$ be n points in G. Let ρ be a positive number and set

$$\omega_i = B(x_i, \rho),$$
$$\Omega = G \setminus \bigcup_{j=1}^{n} \omega_j.$$

Let $d_i \in \mathbb{Z}$, for $i = 1, 2, \ldots, n$, be given and set $d = \sum_{i=1}^{n} d_i$.

We assume

$$\text{(1)} \qquad \operatorname{dist}(x_i, \partial G) \geq 2\mu, \qquad i = 1, 2, \ldots, n, \text{ for some } \mu > 0,$$

$$\text{(2)} \qquad \rho \leq \frac{1}{2} \operatorname{dist}(x_i, \partial G) \qquad i = 1, 2, \ldots, n,$$

and

$$\text{(3)} \qquad 8\rho \leq |x_i - x_j| \qquad \text{for all } i \neq j.$$

The main purpose in this chapter is to provide a lower bound for the energy of maps $v: \Omega \to S^1$ in terms of their degrees on $\partial\omega_i$. In view of the results of Chapter I it is natural to introduce the solution Φ of the problem:

$$\text{(4)} \qquad \begin{cases} \Delta\Phi = 0 & \text{in } \Omega \\ \Phi = \text{Const.} = C_j & \text{on } \partial\omega_j,\ j = 1, 2, \ldots, n, \\ \displaystyle\int_{\partial\omega_j} \frac{\partial\Phi}{\partial\nu} = 2\pi d_j, & j = 1, 2, \ldots, n \\ \Phi = 0 & \text{on } \partial G \end{cases}$$

where ν denotes the outward unit normal to ω_i.

Theorem II.1. *Under the assumptions (1), (2), (3) and also $d > 0$ we have*

$$\int_\Omega |\nabla\Phi|^2 \geq 2\pi \text{ Min}\left(\sum_{j\in P} \delta_j^2\right) \log(\mu/\rho) - C \tag{5}$$

where $P = \{j \in \{1,2,\dots,n\};\ d_j > 0\}$ and the minimum in (5) is taken over all choices of integers δ_j such that $0 \leq \delta_j \leq d_j$ and $\sum_{j\in P} \delta_j = d$, and

$$C = 6\pi\left(\sum_{i=1}^{n} |d_i|\right)^2 \left(\log\left(\frac{\operatorname{diam} G}{\mu}\right) + n \log 2\right).$$

Before proving Theorem II.1 we derive some easy consequences for the energy of the $\mathbf{S}^1$-valued maps. Consider, as in (I.1), the class

$$\mathcal{E} = \left\{ v \in H^1(\Omega; S^1) \,\middle|\, \begin{matrix} \deg(v, \partial G) = d & \text{and} \\ \deg(v, \partial\omega_i) = d_i & i = 1,2,\dots,n \end{matrix} \right\}.$$

Corollary II.1. *Assume (1), (2), (3). Then, for every $v \in \mathcal{E}$, we have*

$$\int_\Omega |\nabla v|^2 \geq 2\pi |d| \log(\mu/\rho) - C$$

where C depends only on $\sum |d_i|$, diam $G/\mu$and n.

Proof of Corollary II.1. Without loss of generality we may assume that $d > 0$. From Theorem I.1 we know that

$$\int_\Omega |\nabla v|^2 \geq \int_\Omega |\nabla\Phi|^2$$

where Φ is the solution of (4). We may then apply Theorem II.1 and note that

$$\delta_j^2 \geq \delta_j$$

to infer that

$$\sum_{j\in P} \delta_j^2 \geq \sum_{j\in P} \delta_j = d.$$

Corollary II.2. *Assume (1), (2), (3) and also*

$$d_j \geq 0 \qquad \forall j = 1, 2, \ldots, n.$$

Then, for every $v \in \mathcal{E}$, we have

$$\int_\Omega |\nabla v|^2 \geq 2\pi \left(\sum_{j=1}^n d_j^2 \right) \log(\mu/\rho) - C$$

where C depends only on $d, \operatorname{diam} G/\mu$ and n.

Proof of Corollary II.2. Consider $P = \{j \in \{1, 2, \ldots n\}; d_j > 0\}$. Note that $\sum_{j \in P} d_j = d$ and therefore the only choice in Theorem II.1 for δ_j is $\delta_j = d_j \quad \forall j \in P$. Combining Theorem I.1 and Theorem II.2 we see that

$$\int_\Omega |\nabla v|^2 \geq 2\pi \left(\sum_{j \in P} d_j^2 \right) \log(\mu/\rho) - C$$

and the conclusion follows since $\sum\limits_{j \in P} d_j^2 = \sum\limits_{j=1}^n d_j^2$.

The proof of Theorem II.1 relies on the following lemma which concerns the solution φ_i of the problem:

$$(6) \qquad \begin{cases} \Delta \varphi_i = 0 & \text{in } \Omega \\ \varphi_i = \text{Const.} & \text{on each } \partial\omega_j, j = 1, 2, \ldots, n, \\ \displaystyle\int_{\partial\omega_j} \frac{\partial \varphi_i}{\partial \nu} = \begin{cases} 2\pi\, d_i \\ 0 \end{cases} & \begin{matrix} \text{if } j = i \\ \text{if } j \neq i \end{matrix} \\ \varphi_i = 0 & \text{on } \partial G. \end{cases}$$

Lemma II.1. *Let φ_i be the solution of (6). Then*

$$(7) \qquad \varphi_i(x) = d_i \log(|x - x_i|/\mu) + d_i\, R_i(x)$$

with, for every $x \in \Omega$,

$$(8) \qquad |R_i(x)| \leq \log\left(\frac{\operatorname{diam} G}{\mu} \right) + (n-1) \log 2 \equiv A.$$

Moreover

$$(9) \qquad \int_\Omega |\nabla \varphi_i|^2 \geq 2\pi\, d_i^2 \log(\mu/\rho) - 2\pi\, d_i^2\, A.$$

Proof of Lemma II.1. We may always assume that $d_i = 1$. Applying Lemma I.4 with

$$v(x) = \varphi_i(x) - \log(|x - x_i|/\mu)$$

we obtain

$$\text{(10)} \qquad \operatorname*{Sup}_{\Omega} v - \operatorname*{Inf}_{\Omega} v \leq \sum_{j=1}^{n} \left(\operatorname*{Sup}_{\partial\omega_j} v - \operatorname*{Inf}_{\partial\omega_j} v \right) + \operatorname*{Sup}_{\partial G} v - \operatorname*{Inf}_{\partial G} v.$$

Set

$$X = \sum_{j=1}^{n} \left(\operatorname*{Sup}_{\partial\omega_j} v - \operatorname*{Inf}_{\partial\omega_j} v \right).$$

Note that for $j \neq i$

$$\begin{aligned} \operatorname*{Sup}_{\partial\omega_j} v - \operatorname*{Inf}_{\partial\omega_j} v &= -\operatorname*{Inf}_{\partial\omega_j} \log|x - x_i| + \operatorname*{Sup}_{\partial\omega_j} \log|x - x_i| \\ &\leq \log \left(\frac{|x_i - x_j| + \rho}{|x_i - x_j| - \rho} \right) \leq \log 2 \end{aligned}$$

by assumption (3).

Also

$$\operatorname*{Sup}_{\partial\omega_i} v - \operatorname*{Inf}_{\partial\omega_i} v = 0 \qquad \text{since } v \text{ is constant on } \partial\omega_i.$$

Thus

$$X \leq (n-1) \log 2.$$

Finally, we have, by (10),

$$\begin{aligned} \operatorname*{Sup}_{\Omega} v &\leq \operatorname*{Inf}_{\Omega} v + \operatorname*{Sup}_{\partial G} v - \operatorname*{Inf}_{\partial G} v + X \\ &\leq \operatorname*{Inf}_{\partial G} v + \operatorname*{Sup}_{\partial G} v - \operatorname*{Inf}_{\partial G} v + X \\ &= \operatorname*{Sup}_{\partial G} v + X. \end{aligned}$$

But

$$\operatorname*{Sup}_{\partial G} v = -\operatorname*{Inf}_{x \in \partial G} \log \left(\frac{|x - x_i|}{\mu} \right) \leq 0 \quad \text{by (1).}$$

Thus

$$\operatorname*{Sup}_{\Omega} v \leq X.$$

Similarly

$$\operatorname*{Inf}_{\Omega} v \geq \operatorname*{Inf}_{\partial G} v - X \geq -\log \left(\frac{\operatorname{diam} G}{\mu} \right) - X.$$

Hence

$$\|v\|_{L^\infty(\Omega)} \le X + \log\left(\frac{\operatorname{diam} G}{\mu}\right).$$

This proves (8).

We now prove (9). We have

$$\int_\Omega |\nabla \varphi_i|^2 = -\sum_{j=1}^n \int_{\partial\omega_j} \varphi_i \frac{\partial \varphi_i}{\partial \nu} = -2\pi\, \varphi_i(\partial\omega_i).$$

Applying (7) and (8) we have

$$|\varphi_i(\partial\omega_i) - \log(\rho/\mu)| \le A$$

and thus

$$\int_\Omega |\nabla\varphi_i|^2 \ge 2\pi\, \log(\mu/\rho) - 2\pi A.$$

Lemma II.2. *Let φ_i be the solution of (6). Assume $d_i > 0$. Then*

(11) $$\varphi_i \le 0 \quad \text{in } \Omega$$

and

(12) $$\frac{\partial \varphi_i}{\partial \nu} \ge 0 \quad \text{on } \partial\omega_i.$$

Proof. We have

$$\begin{aligned} 0 = -\int_\Omega (\Delta\varphi_i)\varphi_i^+ &= \int_\Omega |\nabla\varphi_i^+|^2 + \int_{\partial\omega_i} \frac{\partial\varphi_i}{\partial\nu}\varphi_i^+ \\ &= \int_\Omega |\nabla\varphi_i^+|^2 + 2\pi\, d_i\, \varphi_i^+(\partial\omega_i) \ge \int_\Omega |\nabla\varphi_i^+|^2. \end{aligned}$$

Thus $\varphi_i^+ = 0$ in Ω and hence $\varphi_i \le 0$ in Ω.

We now turn to the proof of (12). Set $C_i = \varphi_i(\partial\omega_i)$. We claim that

(13) $$\varphi_i \ge C_i \quad \text{on } \Omega.$$

This clearly implies (12).

We have

$$0 = -\int_\Omega \Delta(C_i - \varphi_i)(C_i - \varphi_i)^+ = \int_\Omega |\nabla(C_i - \varphi_i)^+|^2 +$$
$$\sum_{j=1}^n \int_{\partial\omega_j} \frac{\partial}{\partial\nu}(C_i - \varphi_i)(C_i - \varphi_i)^+ - \int_{\partial G} \frac{\partial}{\partial\nu}(C_i - \varphi_i)(C_i - \varphi_i)^+.$$

Note that all boundary integrals vanish. Indeed if $j \ne i$, then $(C_i - \varphi_i)^+$ is constant on $\partial\omega_j$ and $\int_{\partial\omega_j} \frac{\partial\varphi_i}{\partial\nu} = 0$. On $\partial\omega_i$ $(C_i - \varphi_i) = 0$, while on ∂G, $(C_i - \varphi_i)^+ = C_i^+ = 0$ by (11). Therefore $(C_i - \varphi_i)^+ = 0$ in Ω and this proves (13).

Lemma II.3. *(Partition "molecules-ions"). Let M be a metric space with:*

(i) k points $n_1, n_2, \ldots, n_k$ (called the -1 points),

(ii) $(k+d)$ points $p_1, p_2, \ldots, p_k, p_{k+1}, \ldots, p_{k+d}$ (called the $+1$ points).

We assume that $d(p_i, n_j) > 0 \ \forall i, j$, but we may possibly have $n_i = n_j$ or $p_i = p_j$ if $i \neq j$. In other words, the -1 points and the $+1$ points are repeated according to their multiplicity.

Then there exists a permutation of the $+1$ points, which are relabelled as,

$$\pi_1, \pi_2, \ldots, \pi_k, P_1, P_2, \ldots, P_d$$

such that

$$\frac{d(P_j, n_i)}{d(P_j, \pi_i)} \geq \frac{1}{2} \quad \forall j = 1, 2, \ldots, d, \quad \forall i = 1, 2, \ldots, k \tag{14}$$

(with possibly $d(P_j, \pi_i) = 0$).

Proof. Starting with n_1, there is a $+1$ point, say p_i, such that

$$d(p_i, n_1) = \operatorname*{Min}_{1 \leq \alpha \leq k+d} d(p_\alpha, n_1). \tag{15}$$

Set

$$\pi_1 = p_i.$$

We have

$$\frac{d(p_j, n_1)}{d(p_j, \pi_1)} \geq \frac{1}{2} \quad \forall j = 1, 2, \ldots, k+d. \tag{16}$$

For, if not, there would be some j, $1 \leq j \leq k+d$, such that

$$d(p_j, n_1) < \frac{1}{2} d(p_j, \pi_1) \leq \frac{1}{2}[d(p_j, n_1) + d(n_1, \pi_1)].$$

Thus

$$d(p_j, n_1) < d(n_1, \pi_1)$$

which contradicts (15). Thus we have proved (16).

Next, we eliminate the pair $\{n_1, \pi_1\}$ and we reiterate the same procedure with n_2. This yields some $+1$ point denoted $\pi_2 \in \{p_1, p_2, \ldots, p_{k+d}\} \backslash \{p_i\}$, such that

$$\frac{d(p_j, n_2)}{d(p_j, \pi_2)} \geq \frac{1}{2} \qquad \forall j \neq i.$$

Then we start with n_3 and so on. When we have exhausted all -1 points we are left with $+1$ points denoted $P_1, P_2, \dots, P_d$. By construction, we see immediately that (14) holds.

Proof of Theorem II.1. Consider the points (x_i) with their associated integers (d_i). We say that a point x_i is a **negative** point if $d_i < 0$; the negative points are repeated according to their multiplicity $|d_i|$ and we denote them by

$$n_1, n_2, \dots, n_k \quad \text{with } k = \sum_{d_i<0} |d_i|.$$

We say that x_i is a **positive** point if $d_i > 0$; the positive points are repeated according to their multiplicity d_i and we denote them by

$$p_1, p_2, \dots, p_{k+d} \quad \text{with } k + d = \sum_{d_i>0} d_i$$

(note that $d = \sum_{d_i>0} d_i - \sum_{d_i<0} |d_i| = \sum_{i=1}^{n} d_i$). The remaining points x_i, associated to $d_i = 0$, are denoted

$$q_1, q_2, \dots, q_r.$$

We relabel the positive points, using Lemma II.3, as

$$\pi_1, \pi_2, \dots, \pi_k, \ P_1, P_2, \dots, P_d$$

in such a way that

$$\frac{d(P_j, n_i)}{d(P_j, \pi_i)} \geq \frac{1}{2} \quad \forall j = 1, 2, \dots, d, \quad \forall i = 1, 2, \dots, k.$$

We say that

$$(\pi_1, n_1), (\pi_2, n_2), \dots, (\pi_k, n_k)$$

are "neutral molecules" and that

$$P_1, P_2, \dots, P_d$$

are "positive ions". We regroup the positive ions according to their multiplicity and write them as

$$[P_1, \delta_1], \ [P_2, \delta_2], \dots, [P_\ell, \delta_\ell]$$

with $\sum_{i=1}^{\ell} \delta_i = d$. We now define corresponding functions

$$\begin{aligned} (\pi_i, n_i) &\longleftrightarrow \psi_i \\ [P_i, \delta_i] &\longleftrightarrow \eta_i \\ q_i &\longleftrightarrow \theta_i \end{aligned}$$

in the following way:

(i) The function ψ_i associated to the neutral molecule (π_i, n_i) is the solution of

$$\begin{aligned} \Delta\psi_i &= 0 && \text{in } \Omega \\ \psi_i &= \text{Const.} && \text{on each } \partial\omega_j,\ j = 1, 2, \ldots, n \\ \psi_i &= 0 && \text{on } \partial G \end{aligned}$$

and

$$\int_{\partial\omega} \frac{\partial\psi_i}{\partial\nu} = \begin{cases} 2\pi & \text{if } \omega \text{ is the disc centered at } \pi_i, \\ -2\pi & \text{if } \omega \text{ is the disc centered at } n_i, \\ 0 & \text{otherwise .} \end{cases}$$

(ii) The function η_i associated to the ion $[P_i, \delta_i]$ is the solution of

$$\begin{aligned} \Delta\eta_i &= 0 && \text{in } \Omega, \\ \eta_i &= \text{Const.} && \text{on each } \partial\omega_j,\ j = 1, 2, \ldots, n, \\ \eta_i &= 0 && \text{on } \partial G, \end{aligned}$$

and

$$\int_{\partial\omega} \frac{\partial\eta_i}{\partial\nu} = \begin{cases} 2\pi\delta_i & \text{if } \omega \text{ is the disc centered at } P_i, \\ 0 & \text{otherwise.} \end{cases}$$

(iii) The function θ_i associated to q_i is the solution of

$$\begin{aligned} \Delta\theta_i &= 0 && \text{in } \Omega \\ \theta_i &= \text{Const.} && \text{on each } \partial\omega_j,\ j = 1, 2, \ldots, n, \\ \theta_i &= 0 && \text{on } \partial G \end{aligned}$$

and

$$\int_{\partial\omega_j} \frac{\partial\theta_i}{\partial\nu} = 0 \qquad \forall j = 1, 2, \ldots, n.$$

So that, in fact, $\theta_i \equiv 0$.

By construction, we obtain

$$\Phi = \sum_{i=1}^{k} \psi_i + \sum_{j=1}^{\ell} \eta_j.$$

Hence

$$(17) \qquad \begin{cases} |\nabla\Phi|^2 & = |\sum_{i=1}^{k} \nabla\psi_i|^2 + |\sum_{j=1}^{\ell} \nabla\eta_j|^2 + 2\sum_{i,j} \nabla\psi_i \nabla\eta_j \\ & \geq |\sum_{j=1}^{\ell} \nabla\eta_j|^2 + 2\sum_{i,j} \nabla\psi_i \nabla\eta_j. \end{cases}$$

Step 1: We have

$$(18) \qquad \int_\Omega |\sum_{j=1}^{\ell} \nabla\eta_j|^2 \geq 2\pi \sum_{j=1}^{\ell} \delta_j^2 \log(\mu/\rho) - 2\pi A \sum_{j=1}^{\ell} \delta_j^2$$

where A is defined in Lemma II.1.

Proof of (18). Write

$$\int_\Omega |\sum_{j=1}^{\ell} \nabla\eta_j|^2 = \sum_{j=1}^{\ell} \int_\Omega |\nabla\eta_j|^2 + 2\sum_{j\neq s} \int_\Omega \nabla\eta_j \nabla\eta_s.$$

But

$$\begin{aligned} \int_\Omega \nabla\eta_j \nabla\eta_s &= -\sum_{m=1}^{n} \int_{\partial\omega_m} \eta_j \frac{\partial\eta_s}{\partial\nu} = -\sum_{m=1}^{n} \eta_j(\partial\omega_m) \int_{\partial\omega_m} \frac{\partial\eta_s}{\partial\nu} \\ &= -\eta_j(\partial\omega)\, 2\pi\delta_s \end{aligned}$$

where ω is the disc centered at P_s. By Lemma II.2 we know that $\eta_j \leq 0$ in Ω and thus $\int_\Omega \nabla\eta_j \nabla\eta_s \geq 0$.

On the other hand, we deduce from Lemma II.1 that

$$\int_\Omega |\nabla\eta_j|^2 \geq 2\pi\delta_j^2 \log(\mu/\rho) - 2\pi A\, \delta_j^2.$$

This proves (18).

Step 2: We have $\forall i, j$

$$(19) \qquad \int_\Omega \nabla\psi_i \nabla\eta_j \geq -2\pi\delta_j(2A + \log 4).$$

Proof of (19). Write

$$\int_\Omega \nabla\psi_i \nabla\eta_j = -\sum_{m=1}^{n} \int_{\partial\omega_m} \psi_i \frac{\partial\eta_j}{\partial\nu} = -\psi_i(\partial\omega_j)2\pi\delta_j, \tag{20}$$

where $\omega_j = B(P_j, \rho)$. Recall that ψ_i is associated to the neutral molecule (π_i, n_i). By Lemma II.1 we have, in Ω,

$$|\psi_i(x) - \log\left(\frac{|x-\pi_i|}{|x-n_i|}\right)| \le 2A. \tag{21}$$

Two cases may occur:

Case 1: $\pi_i = P_j$.

Case 2: $\pi_i \neq P_j$.

Case 1. On $\partial\omega_j$ we have

$$\psi_i(x) \le \log\left(\frac{\rho}{|x-n_i|}\right) + 2A$$

and therefore

$$\psi_i(\partial\omega_j) \le \log\left(\frac{\rho}{|n_i - P_j| + \rho}\right) + 2A \le 2A.$$

Case 2. On $\partial\omega_j$ we have

$$\begin{aligned}\psi_i(x) &\le \log\left(\frac{|\pi_i - P_j| + \rho}{|n_i - P_j| - \rho}\right) + 2A \\ &= \log\left(\frac{|\pi_i - P_j|}{|n_i - P_j|}\right) + \log\left(\frac{1 + (\rho/|\pi_i - P_j|)}{1 - (\rho/|n_i - P_j|)}\right) + 2A \\ &\le \log\left(\frac{|\pi_i - P_j|}{|n_i - P_j|}\right) + \log 2 + 2A, \qquad \text{by (3).}\end{aligned}$$

By Lemma II.3 we have

$$\frac{|\pi_i - P_j|}{|n_i - P_j|} \le 2.$$

Therefore we finally obtain, for $x \in \partial\omega_j$,

$$\psi_i(x) \le \log 4 + 2A.$$

In both Case 1 and Case 2 we conclude that

$$\psi_i(\partial\omega_j) \le \log 4 + 2A.$$

Going back to (20) we find

$$\int_\Omega \nabla\psi_i \nabla\eta_j \geq -2\pi\delta_j(2A + \log 4).$$

Thus Step 2 is proved.

Proof of Theorem II.1 completed. Combining Step 1, Step 2 and (17) we are led to

$$\int_\Omega |\nabla\Phi|^2 \geq 2\pi \sum_{j=1}^{\ell} \delta_j^2 \log(\mu/\rho) - 2\pi A \sum_{j=1}^{\ell} \delta_j^2 - 2\pi k(2A + \log 4) \sum_{j=1}^{\ell} \delta_j. \tag{22}$$

Note that

$$\sum_{j=1}^{\ell} \delta_j^2 \leq \left(\sum_{j=1}^{\ell} \delta_j\right)^2$$

and

$$k \leq \sum_{i=1}^{n} |d_i|.$$

Finally we obtain

$$\int_\Omega |\nabla\Phi|^2 \geq 2\pi \sum_{j=1}^{\ell} \delta_j^2 \log(\mu/\rho) - 2\pi(3A + \log 4)\left(\sum_{i=1}^{n} |d_i|\right)^2.$$

CHAPTER III

Some basic estimates for u_ε

Let $G \subset \mathbb{R}^2$ be a smooth, bounded and simply connected domain. Consider the functional

$$E_\varepsilon(u) = \frac{1}{2}\int_G |\nabla u|^2 + \frac{1}{4\varepsilon^2}\int_G (|u|^2 - 1)^2 \tag{1}$$

which is defined for maps $u \in H^1(G;\mathbb{C})$ and let

$$H^1_g = \{u \in H^1(G;\mathbb{C}); u = g \text{ on } \partial G\}$$

where $g : \partial G \to \mathbb{C}$ is a prescribed smooth map such that $|g(x)| = 1$ $\forall\, x \in \partial\Omega$.

Throughout the rest of the book we assume that $d = \deg(g, \partial\Omega) > 0$.

It is easy to see that

$$\min_{u \in H^1_g} E_\varepsilon(u) \tag{2}$$

is achieved by some u_ε that satisfies the Euler equation

$$\begin{cases} -\Delta u_\varepsilon = \dfrac{1}{\varepsilon^2} u_\varepsilon(1 - |u_\varepsilon|^2) & \text{in } G, \\ u_\varepsilon = g & \text{on } \partial G. \end{cases} \tag{3}$$

The maximum principle implies (see e.g., F. Bethuel, H. Brezis and F. Hélein [2]) that any solution u_ε of (3) satisfies

$$|u_\varepsilon| \le 1 \quad \text{in } G. \tag{4}$$

III.1. Estimates when $G = B_R$ and $g(x) = x/|x|$

The following quantity will play an important role. Given $\varepsilon > 0$ and $R > 0$ set

$$I(\varepsilon, R) = \min_{u \in H^1_g} \left\{ \frac{1}{2}\int_{B_R} |\nabla u|^2 + \frac{1}{4\varepsilon^2}\int_{B_R} (|u|^2 - 1)^2 \right\} \tag{5}$$

where $g(x) = x/|x|$ on ∂B_R and

$$I(t) = I(t,1) \quad \text{for } t > 0. \tag{6}$$

By scaling it is easy to see that

$$I(\varepsilon, R) = I\left(\frac{\varepsilon}{R}\right) = I\left(1, \frac{R}{\varepsilon}\right). \tag{7}$$

Remark III.1. Note that $I(t) \to +\infty$ as $t \to 0$. Suppose not; say that there is a sequence $t_n \to 0$ such that $I(t_n) \le C$. By choosing a minimizer in (5) we would have a sequence $u_n : B_1 \to \mathbb{C}$ such that

$$\int_{B_1} |\nabla u_n|^2 \le C,$$

$$\int_{B_1} (|u_n|^2 - 1)^2 \le C\, t_n^2$$

and

$$u_n(x) = x \quad \text{on } \partial B_1.$$

Passing to a subsequence we would find some $u \in H^1(B_1; S^1)$ such that $u(x) = x$ on ∂B_1. This is impossible (see the Introduction).

Lemma III.1. *We have,* $\forall t_1 \le t_2$,

$$I(t_1) \le \pi \log(t_2/t_1) + I(t_2)$$

i.e., the function $t \mapsto (I(t) + \pi \log t)$ *is nondecreasing.*

In particular

$$I(t) \le \pi \log(1/t) + I(1) \quad \forall t \in (0,1].$$

Proof. Let u_2 be a minimizer for $I(t_2) = I\left(1, \frac{1}{t_2}\right)$.

Set

$$u_1(x) = \begin{cases} u_2(x) & \text{if } |x| < \dfrac{1}{t_2}, \\[2mm] \dfrac{x}{|x|} & \text{if } \dfrac{1}{t_2} < |x| < \dfrac{1}{t_1}. \end{cases}$$

We have

$$I(t_1) = I\left(1, \frac{1}{t_1}\right) \le \frac{1}{2}\int_{B_{1/t_1}} |\nabla u_1|^2 + \frac{1}{4}\int_{B_{1/t_1}} (|u_1|^2 - 1)^2$$

$$= \frac{1}{2}\int_{B_{1/t_2}} |\nabla u_2|^2 + \frac{1}{4}\int_{B_{1/t_2}} (|u_2|^2 - 1)^2 + \frac{1}{2}\int_{B_{1/t_1}\setminus B_{1/t_2}} \left|\nabla\left(\frac{x}{|x|}\right)\right|^2$$

$$= I(t_2) + \pi \log(t_2/t_1).$$

Remark III.2 One may also prove that there is a universal constant C such that

$$I(t) \geq \pi \log(1/t) - C \quad \forall t \in (0,1]. \tag{8}$$

We do not have a simple proof of (8)—see Theorem V.3.

III.2. An upper bound for $E_\varepsilon(u_\varepsilon)$

Let u_ε be a minimizer for (1).

Theorem III.1. *We have, for $\varepsilon < \varepsilon_0$,*

$$E_\varepsilon(u_\varepsilon) \leq \pi d \log(1/\varepsilon) + C \tag{9}$$

where ε_0 and C depend only on g and G.

Proof. Fix d distinct points $a_1, a_2, \ldots, a_d$ in G and fix $R > 0$ so small that

$$\overline{B(a_i, R)} \subset G \quad \forall i \quad \text{and} \quad \overline{B(a_i, R)} \cap \overline{B(a_j, R)} = \emptyset \qquad \forall i \neq j.$$

Let $\Omega = G \backslash \left(\bigcup_{i=1}^d \overline{B(a_i, R)} \right)$ and consider the map $\bar{g} : \partial\Omega \to S^1$ defined by

$$\bar{g}(x) = \begin{cases} g(x) & \text{if } \ x \in \partial G, \\ e^{i\theta} & \text{if } \ x = a_j + Re^{i\theta} \in \partial B(a_j, R). \end{cases}$$

Since

$$\deg(\bar{g}, \partial\Omega) = 0$$

there is a smooth map $\bar{v} : \bar{\Omega} \to S^1$ such that $\bar{v} = \bar{g}$ on $\partial\Omega$.

We have, by Lemma III.1, for $\varepsilon < R$,

$$E_\varepsilon(u_\varepsilon) \leq \frac{1}{2} \int_\Omega |\nabla \bar{v}|^2 + \sum_{i=1}^d I(\varepsilon, R) \leq \pi d \log(1/\varepsilon) + C$$

which is the desired estimate.

III.3. An upper bound for $\frac{1}{\varepsilon^2}\int_G(|u_\varepsilon|^2-1)^2$

Theorem III.2. *Assume G is starshaped about the origin (i.e., $x\cdot\nu\geq\alpha>0$ $\forall x\in\partial G$). Then there is a constant C depending only on g and G such that any solution u_ε of (3) satisfies*

$$\int_{\partial G}\left|\frac{\partial u_\varepsilon}{\partial\nu}\right|^2+\frac{1}{\varepsilon^2}\int_G(|u_\varepsilon|^2-1)^2\leq C. \tag{10}$$

Estimate (10) plays a crucial role in our analysis. **Therefore we assume throughout the rest of the book that G is starshaped.**

Proof. As in the proof of the Pohozaev identity one multiplies (3) by $x\cdot\nabla u_\varepsilon=x_1\dfrac{\partial u_\varepsilon}{\partial x_1}+x_2\dfrac{\partial u_\varepsilon}{\partial x_2}$.

This yields, dropping the subscript ε,

$$\begin{aligned}-\int_{\partial G}\frac{\partial u}{\partial\nu}(x\cdot\nabla u)+\int_G\sum_{i,j}u_{x_j}\left(\delta_{ij}u_{x_i}+x_iu_{x_ix_j}\right)&\\=\frac{1}{2\varepsilon^2}\int_G(|u|^2-1)^2.&\end{aligned}\tag{11}$$

But

$$\frac{\partial u}{\partial\nu}(x\cdot\nabla u)=(x\cdot\nu)\left(\frac{\partial u}{\partial\nu}\right)^2+(x\cdot\tau)\frac{\partial u}{\partial\tau}\frac{\partial u}{\partial\nu}\tag{12}$$

and

$$\begin{aligned}\int_G\sum_{i,j}u_{x_j}(x_iu_{x_ix_j})&=\frac12\int_G\sum_i x_i\frac{\partial}{\partial x_i}|\nabla u|^2\\&=-\int_G|\nabla u|^2+\frac12\int_{\partial G}(x\cdot\nu)|\nabla u|^2.\end{aligned}\tag{13}$$

Combining (11), (12) and (13) we obtain

$$\begin{aligned}\frac12\int_{\partial G}(x\cdot\nu)\left(\frac{\partial u}{\partial\nu}\right)^2+\frac{1}{2\varepsilon^2}\int_G(|u|^2-1)^2&\\=\int_{\partial G}\frac12(x\cdot\nu)\left(\frac{\partial g}{\partial\tau}\right)^2-(x\cdot\tau)\frac{\partial u}{\partial\nu}\frac{\partial g}{\partial\tau}.&\end{aligned}$$

This directly implies (10).

III.4. $|u_\varepsilon| \geq 1/2$ on "good discs"

Theorem III.3. *There exist positive constants λ_0 and μ_0 (depending only on G and g) such that if u_ε is a solution of (3) satisfying*

$$\frac{1}{\varepsilon^2}\int_{G\cap B_{2\ell}} (|u_\varepsilon|^2-1)^2 \leq \mu_0 \tag{14}$$

where $B_{2\ell}$ is some disc of radius 2ℓ in $\mathbb{R}^2$ with

$$\frac{\ell}{\varepsilon} \geq \lambda_0 \quad \text{and } \ell \leq 1, \tag{15}$$

then

$$|u_\varepsilon(x)| \geq \frac{1}{2} \quad \forall x \in G \cap B_\ell. \tag{16}$$

Proof. It follows from Lemma A.2 in the Appendix of F. Bethuel, H. Brezis and F. Hélein [2] that

$$\|\nabla u_\varepsilon\|_{L^\infty(G)} \leq C/\varepsilon,$$

where C depends only on G and g. Therefore, we have

$$|u_\varepsilon(x) - u_\varepsilon(y)| \leq \frac{C}{\varepsilon}|x-y| \quad \forall x,\, y \in G. \tag{17}$$

We argue by contradiction and assume that $|u_\varepsilon(x_0)| < 1/2$ for some $x_0 \in G \cap B_\ell$. Then we have

$$|u_\varepsilon(x) - u_\varepsilon(x_0)| \leq \frac{C}{\varepsilon}|x-x_0|$$

and thus

$$|u_\varepsilon(x)| \leq \frac{1}{2} + \frac{C}{\varepsilon}\rho \quad \text{in } G\cap B_\rho(x_0).$$

Consequently

$$1 - |u_\varepsilon(x)| \geq \frac{1}{2} - \frac{C}{\varepsilon}\rho \quad \text{in } G\cap B_\rho(x_0).$$

We choose $\rho = \varepsilon/4C$, so that

$$1 - |u_\varepsilon(x)| \geq \frac{1}{4} \quad \text{in } G\cap B_{\varepsilon/4C}(x_0)$$

and consequently

$$(|u_\varepsilon(x)|^2 - 1)^2 \geq \frac{1}{16} \quad \text{in } G\cap B_{\varepsilon/4C}(x_0).$$

On the other hand there is a positive constant α (depending only on G), such that

$$\text{meas}\,(G \cap B_r(x)) \geq \alpha r^2 \quad \forall x \in G,\ \forall r \leq 1.$$

Hence, we have

$$\int_{G \cap B_{\varepsilon/4C}(x_0)} (|u_\varepsilon|^2 - 1)^2 \geq \frac{\alpha\varepsilon^2}{(16C)^2}$$

provided $(\varepsilon/4C) \leq 1$. Note that $B_{\varepsilon/4C}(x_0) \subset B_{2\ell}$ when $(\varepsilon/4C) \leq \ell$ (since $x_0 \in B_\ell$).

Therefore

$$\int_{G \cap B_{2\ell}} (|u_\varepsilon|^2 - 1)^2 \geq \frac{\alpha\varepsilon^2}{(16C)^2}.$$

If we choose $\lambda_0 = 1/4C$ and $\mu_0 < \alpha/(16C)^2$ we are led to a contradiction.

CHAPTER IV

Towards locating the singularities: bad discs and good discs

For some technical reasons it is convenient to enlarge a little the given domain G. Fix a smooth, bounded and simply connected domain G' such that $\bar{G} \subset G'$. Also fix arbitrarily a smooth map $\bar{g} : G' \backslash G \to S^1$ such that

$$\bar{g} = g \quad \text{on } \partial G.$$

Clearly, such a construction is always possible. We extend systematically any map $u : G \to \mathbb{C}$ with $u = g$ on ∂G, by a map, still denoted u, $u : G' \to \mathbb{C}$, such that $u = \bar{g}$ in $G' \backslash G$. In particular, u_ε introduced in Chapter III is now defined on G'.

IV.1. A covering argument

Let u_ε be a minimizer for E_ε in H^1_g. We shall now use a covering argument that helps to locate the (small) regions where u_ε has a singular behavior.

Consider a family of discs $B(x_i, \lambda_0\varepsilon)_{i\in I}$, where λ_0 is defined in Theorem III.3, such that

$$x_i \in G, \quad \forall i \in I, \tag{1}$$

$$B(x_i, \lambda_0\varepsilon/4) \cap B(x_j, \lambda_0\varepsilon/4) = \emptyset, \quad \forall i \neq j, \tag{2}$$

$$\bigcup_{i\in I} B(x_i, \lambda_0\varepsilon) \supset G. \tag{3}$$

For this purpose, it suffices to consider a maximal family satisfying (1) and (2).

We say that the disc $B(x_i, \lambda_0\varepsilon)$ is a **good disc** if

$$\frac{1}{\varepsilon^2}\int_{B(x_i, 2\lambda_0\varepsilon)} (|u_\varepsilon|^2 - 1)^2 < \mu_0$$

where μ_0 is defined in Theorem III.3.

$B(x_i, \lambda_0\varepsilon)$ is a bad disc if

$$\frac{1}{\varepsilon^2}\int_{B(x_i,2\lambda_0\varepsilon)}(|u_\varepsilon|^2-1)^2 \geq \mu_0.$$

The collection of bad discs is labelled by

$$J = \{i\,; B(x_i, \lambda_0\varepsilon) \quad \text{is a bad disc}\}.$$

The following lemma plays an essential role:

Lemma IV.1. *There exists an integer N that depends only on g and G such that*

$$\operatorname{card} J \leq N. \tag{4}$$

Remark IV.1. Strictly speaking we should have denoted the points $(x_i)_{i\in J}$ by $(x_i^\varepsilon)_{i\in J_\varepsilon}$. The main content of Lemma IV.1 is that card J_ε remains bounded independently of ε.

Proof. There is a universal constant C such that

$$\sum_{i\in I}\int_{B(x_i,2\lambda_0\varepsilon)}(|u_\varepsilon|^2-1)^2 \leq C\int_G(|u_\varepsilon|^2-1)^2$$

since each point in G is covered by at most C discs $B(x_i, 2\lambda_0\varepsilon)$. It follows that

$$\mu_0 \operatorname{card} J \leq \frac{C}{\varepsilon^2}\int_G(|u_\varepsilon|^2-1)^2$$

and by Theorem III.2. we deduce that (4) holds.

Lemma IV.2. *We have*

$$|u_\varepsilon(x)| \geq \frac{1}{2} \quad \forall x \in G' \setminus \bigcup_{i\in J} B(x_i, \lambda_0\varepsilon).$$

Proof. Let $x \in G\setminus\bigcup_{i\in J} B(x_i, \lambda_0\varepsilon)$. By (3), there is some $j \in I\setminus J$ such that

$$x \in B(x_j, \lambda_0\varepsilon), \quad \text{which is a good disc.}$$

It follows from Theorem III.3. that $|u_\varepsilon(x)| \geq 1/2$.

IV.2 Modifying the bad discs

In this section we shall replace the bad discs $B(x_i, \lambda_0\varepsilon)_{i\in J}$ by slightly larger discs (deleting if necessary some of the points (x_i)) in such a way that the points x_i are far apart (relative to ε).

Theorem IV. 1. *We may choose a subset $J' \subset J$ and a constant $\lambda \geq \lambda_0$, depending only on g and G, such that*

$$|x_i - x_j| \geq 8\lambda\varepsilon\,, \quad \forall i,j \in J', i \neq j \tag{5}$$

and

$$\bigcup_{i\in J} B(x_i, \lambda_0\,\varepsilon) \subset \bigcup_{i\in J'} B(x_i, \lambda\,\varepsilon). \tag{6}$$

Proof. The argument is by induction on card J. If (5) holds with $J' = J$ and $\lambda = \lambda_0$ we are done. Otherwise there is a pair, say x_1, x_2, such that

$$|x_1 - x_2| < 8\lambda_0\,\varepsilon. \tag{7}$$

We take $\lambda = 9\lambda_0$ and $J' = J\backslash\{1\}$. We are reduced to the previous case. After a finite number of steps (at most N) we are led to the conclusion of the theorem with $\lambda_0 \leq \lambda \leq \lambda_0\, 9^{\text{card}\, J}$.

Roughly speaking the points $(x_i)_{i\in J'} = (x_i^\varepsilon)_{i\in J'}$ correspond to points where u_ε may have a singular behavior. To simplify the notation we shall write $J = J_\varepsilon$ instead of J'. We summarize the main properties of the family (x_i):

$$|u_\varepsilon(x)| \geq \frac{1}{2} \qquad \forall x \in G' \backslash \bigcup_{i\in J} B(x_i, \lambda\varepsilon) \tag{8}$$

$$|x_i - x_j| \geq 8\lambda\varepsilon \qquad \forall i, j \in J, i \neq j \tag{9}$$

$$\text{card}\, J \leq N \tag{10}$$

and

$$\frac{1}{\varepsilon^2}\int_{B(x_i,2\lambda\varepsilon)} \left(|u_\varepsilon|^2 - 1\right)^2 \geq \mu_0 \quad \forall i \in J. \tag{11}$$

Given any sequence $\varepsilon_n \to 0$ we may extract a subsequence (still denoted ε_n) such that

$$\text{card}\, J_{\varepsilon_n} \equiv \text{Const.} = N_1 \tag{12}$$

and

$$x_i^{\varepsilon_n} \longrightarrow \ell_i \in \bar{G} \qquad \forall i = 1, 2, \ldots N_1. \tag{13}$$

We cannot exclude the possibility that some of the points ℓ_i are the same, i.e., it may happen that $x_i^{\varepsilon_n}$ and $x_j^{\varepsilon_n}$ converge to the same limit.

We denote by

$$a_1, a_2, \ldots, a_{N_2} \qquad \text{with } N_2 \le N_1$$

the collection of **distinct points** in (ℓ_i). We are going to prove in Chapter V that for every compact subset K of $\bar{G} \setminus \bigcup\{a_i\}$

$$\int_K |\nabla u_{\varepsilon_n}|^2 \quad \text{remains bounded}$$

and this, in turn, will imply that, on K, $u_{\varepsilon_n} \to u_\star$ uniformly. Then we shall prove in Chapter VI that $a_j \notin \partial G$.

CHAPTER V

An upper bound for the energy of u_ε away from the singularities

Fix $\eta > 0$ such that

(1) $$\eta < \operatorname{dist}(G, \partial G')$$

(2) $$\eta < \frac{1}{2}|a_i - a_j| \quad \forall i \neq j$$

so that the discs $B(a_j, \eta)$ are disjoint and contained in G'.

Clearly, we have, for n sufficiently large, say $n \geq N(\eta)$, depending on η,

(3) $$\bigcup_{i \in J} B(x_i^{\varepsilon_n}, \lambda\varepsilon_n) \subset \bigcup_j B(a_j, \eta/4).$$

In what follows we shall often write x_i instead of $x_i^{\varepsilon_n}$.

Recall that, by (IV.8),

$$|u_{\varepsilon_n}(x)| \geq \frac{1}{2} \quad \text{for } x \in \partial B(a_j, \eta/2),\ n \geq N(\eta)$$

and thus

$$\deg(u_{\varepsilon_n}, \partial B(a_j, \eta/2))$$

is well defined and it remains bounded (as $n \to +\infty$) by the following

Lemma V.1. *We have,* $\forall i \in J$,

(4) $$|\deg(u_\varepsilon, \partial B(x_i^\varepsilon, \lambda\varepsilon))| \leq C, \quad \textit{independent of } \varepsilon.$$

Proof. Recall that

$$\deg(u_\varepsilon, \partial B(x_i^\varepsilon, \lambda\varepsilon)) = \frac{1}{2\pi}\int_{\partial B(x_i^\varepsilon, \lambda\varepsilon)} \frac{u_\varepsilon}{|u_\varepsilon|^2} \wedge (u_\varepsilon)_\tau.$$

The conclusion follows from the fact that $|u_\varepsilon| \geq 1/2$ on $\partial B(x_i^\varepsilon, \lambda\varepsilon)$ and $\|\nabla u_\varepsilon\|_{L^\infty} \leq C/\varepsilon$ (see Lemma A.2 in the Appendix of F. Bethuel, H. Brezis and F. Hélein [2]).

Passing to a subsequence we may assume that

$$d_i = \deg(u_{\varepsilon_n}, \partial B(x_i^{\varepsilon_n}, \lambda\varepsilon_n)) \quad \text{is independent of } n.$$

For the same reason

$$\kappa_j = \deg(u_{\varepsilon_n}, \partial B(a_j, \eta/2)) \quad \text{is also independent of } n.$$

The main result in this chapter is:

Theorem V.1. *There exist a constant C (depending only on g and G, but independent of η and n) and an integer $N(\eta)$, such that, for all $n \geq N(\eta)$,*

$$\int_{G\setminus\bigcup_j B(a_j,\eta)} |\nabla u_{\varepsilon_n}|^2 \leq 2\pi d |\log \eta| + C. \tag{5}$$

The proof of this theorem is indirect; it relies on a lower bound for the energy near a_j which is presented in Section V.1. We will return to the proof of Theorem V.1 in Section V.2.

V.1. A lower bound for the energy of u_ε near a_j

For every $j = 1, 2, \ldots, N_2$, we set

$$\Lambda_j = \left\{ i \in \{1, 2, \ldots, N_1\} ;\ x_i^{\varepsilon_n} \to a_j \right\}$$

and consequently

$$\sum_{i\in\Lambda_j} d_i = \kappa_j \quad \forall j = 1, 2, \ldots, N_2.$$

Set

$$\Omega_j = B(a_j, \eta) \setminus \bigcup_{i\in\Lambda_j} B(x_i^{\varepsilon_n}, \lambda\varepsilon_n)$$

(Ω_j also depends on n, but for simplicity we drop n).

An important estimate is given by the following

Theorem V.2. *There exists a constant C (independent of n and η) such that, for every j and every $n \geq N(\eta)$, we have*

$$\int_{\Omega_j} |\nabla u_{\varepsilon_n}|^2 \geq 2\pi|\kappa_j| \log(\eta/\varepsilon_n) - C. \tag{6}$$

Proof. We write on Ω_j

$$u_{\varepsilon_n} = |u_{\varepsilon_n}| v_{\varepsilon_n} \quad \text{where } v_{\varepsilon_n} = u_{\varepsilon_n}/|u_{\varepsilon_n}|.$$

Since v_{ε_n} is S^1-valued and $\deg(v_{\varepsilon_n}, \partial B(x_i^{\varepsilon_n}, \lambda\varepsilon_n)) = d_i$ we know by Corollary II.1 that

$$\int_{\Omega_j} |\nabla v_{\varepsilon_n}|^2 \geq 2\pi|\kappa_j| \log(\eta/\varepsilon_n) - C. \tag{7}$$

On the other hand, we have

$$|\nabla u_{\varepsilon_n}|^2 = |u_{\varepsilon_n}|^2 |\nabla v_{\varepsilon_n}|^2 + \big|\nabla |u_{\varepsilon_n}|\big|^2$$

and therefore

$$(8) \qquad \int_{\Omega_j} |\nabla u_{\varepsilon_n}|^2 \geq \int_{\Omega_j} |\nabla v_{\varepsilon_n}|^2 - \int_{\Omega_j} (1 - |u_{\varepsilon_n}|^2) |\nabla v_{\varepsilon_n}|^2.$$

Next, we claim that

$$(9) \qquad \int_{\Omega_j} (1 - |u_{\varepsilon_n}|^2) |\nabla v_{\varepsilon_n}|^2 \leq C.$$

Using the fact that $|u_{\varepsilon_n}| \geq 1/2$ on Ω_j we see that

$$|\nabla v_{\varepsilon_n}| \leq C |\nabla u_{\varepsilon_n}| \quad \text{on } \Omega_j$$

and therefore, by Cauchy-Schwarz,

$$(10) \qquad \int_{\Omega_j} (1 - |u_{\varepsilon_n}|^2) |\nabla v_{\varepsilon_n}|^2 \leq C \big\|(1 - |u_{\varepsilon_n}|^2)\big\|_2 \, \|\nabla u_{\varepsilon_n}\|_4^2.$$

Recall that (see Theorem III.2)

$$(11) \qquad \big\|(1 - |u_{\varepsilon_n}|^2)\big\|_2 \leq C\,\varepsilon.$$

From the Euler equation (III.3) we have

$$(12) \qquad \|u_\varepsilon\|_{H^2} \leq \frac{1}{\varepsilon^2} \big\|(1 - |u_\varepsilon|^2)\big\|_2 + C \leq C/\varepsilon.$$

Combining (12) and the estimate $\|u_\varepsilon\|_\infty \leq 1$ we derive, with the help of the Gagliardo-Nirenberg inequality, that

$$(13) \qquad \|\nabla u_\varepsilon\|_4 \leq C \|u_\varepsilon\|_{H^2}^{1/2} \|u_\varepsilon\|_\infty^{1/2} \leq C/\varepsilon^{1/2}.$$

Going back to (10) and using (11) with (13) we are led to (9).

Finally, (7), (8) and (9) yield (6).

V.2. Proof of Theorem V.1

Before proving Theorem V.1 we must derive some consequences of Theorem V.2.

Lemma V.2. *We have*

$$\kappa_j \geq 0 \quad \forall j.$$

Proof. Applying Theorem V.2 we see that

$$\int_{\Omega_j} |\nabla u_{\varepsilon_n}|^2 \geq 2\pi|\kappa_j|\,|\log \varepsilon_n| - C(\eta). \tag{14}$$

Hence

$$\sum_j \int_{\Omega_j} |\nabla u_{\varepsilon_n}|^2 \geq 2\pi|\log \varepsilon_n| \sum_j |\kappa_j| - C(\eta). \tag{15}$$

On the other hand, we recall that, by Theorem III.1,

$$\int_G |\nabla u_{\varepsilon_n}|^2 \leq 2\pi\, d|\log \varepsilon_n| + C. \tag{16}$$

Combining (15) and (16) we see that

$$\sum_j |\kappa_j| \leq d + \frac{C(\eta)}{|\log \varepsilon_n|}.$$

As $n \to \infty$ (since κ_j is independent of n) we find

$$\sum_j |\kappa_j| \leq d. \tag{17}$$

On the other hand we have

$$\sum_j \kappa_j = d.$$

It follows that $\kappa_j \geq 0 \quad \forall j.$

Going back to Theorem V.2, in the light of Lemma V.2, we may now state

Theorem V.3. *There exists a constant C, depending on g and G such that*

$$\int_G |\nabla u_\varepsilon|^2 \geq 2\pi\ d|\log \varepsilon| - C, \quad \forall \varepsilon \leq 1. \tag{18}$$

Proof. Suppose, by contradiction, that (18) does not hold. Then there exists a sequence $\varepsilon_n \leq 1$ such that

$$\int_G |\nabla u_{\varepsilon_n}|^2 - 2\pi d|\log \varepsilon_n| \to -\infty. \tag{19}$$

By Theorem V.2 and Lemma V.2 we know that

$$\int_{\Omega_j} |\nabla u_{\varepsilon_n}|^2 \geq 2\pi\kappa_j \log(\eta/\varepsilon_n) - C.$$

Summing over j we obtain

$$\int_{\bigcup_j \Omega_j} |\nabla u_{\varepsilon_n}|^2 \geq 2\pi\, d \log(\eta/\varepsilon_n) - C. \tag{20}$$

This yields a contradiction with (19) since η is fixed.

Proof of Theorem V.1. Combining (20) with the upper bound of Theorem III.1 we see that

$$\int_{G\setminus \bigcup_j B(a_j,\eta)} |\nabla u_{\varepsilon_n}|^2 \leq 2\pi\, d|\log \eta| + C.$$

CHAPTER VI

u_{ε_n} converges: $u_\star$ is born!

To summarize the result of Chapter V we have now found a subsequence (u_{ε_n}) and a finite set (a_j) in $\bar{G}$ such that on every compact subset K of $G' \setminus \bigcup_j \{a_j\}$ we have

$$\int_K |\nabla u_{\varepsilon_n}|^2 \leq C_K \tag{1}$$

and

$$|u_{\varepsilon_n}(x)| \geq \frac{1}{2} \quad \forall x \in G' \setminus K, \quad \forall n \geq n_K. \tag{2}$$

Passing to a further subsequence (and using a standard diagonal procedure) we may now finally (!) assert that

$$u_{\varepsilon_n} \to u_\star \text{ a.e. on } G. \tag{3}$$

Since

$$\int_G (|u_{\varepsilon_n}|^2 - 1)^2 \leq C\varepsilon_n^2$$

we deduce that $|u_\star| = 1$ a.e. We also have

$$u_{\varepsilon_n} \rightharpoonup u_\star \text{ weakly in } H^1(K) \text{ for every } K \text{ as above.} \tag{4}$$

Recall that $u_\star = \bar{g}$ on $G' \setminus G$ (see the beginning of Chapter IV).

The first result in this chapter is

Theorem VI.1. *We have*

$$u_\star \in C^\infty(G \setminus \bigcup_j \{a_j\}; S^1) \tag{5}$$

(6) *$u_\star$ is a harmonic map, i.e.*

$$-\Delta u_\star = u_\star |\nabla u_\star|^2 \quad \text{in } G \setminus \bigcup_j \{a_j\},$$

(7)
$$u_\star = g \quad \text{on } \partial G,$$

(8)
$$\deg(u_\star, a_j) \geq 0 \quad \forall j,$$

(9)
$$\sum_j \deg(u_\star, a_j) = d,$$

and, for every compact subset $K \subset G \setminus \bigcup_j \{a_j\}$, for every integer k,

(10)
$$u_{\varepsilon_n} \to u_\star \quad \text{in } C^k(K),$$

(11)
$$\frac{1 - |u_{\varepsilon_n}|^2}{\varepsilon_n^2} \to |\nabla u_\star|^2 \quad \text{in } C^k(K),$$

(12)
$$u_{\varepsilon_n} \to u_\star \quad \text{in } C^{1,\alpha}_{\text{loc}}\left(\overline{G} \setminus \bigcup_j \{a_j\}\right), \quad \forall \alpha < 1.$$

Remark VI.1. More precisely, we prove that given any sequence $\varepsilon_n \to 0$, there is a subsequence, still denoted ε_n, and a finite set (a_j), and a map $u_\star$, such that all the conclusions of Theorem VI.1 hold.

VI.1. Proof of Theorem VI.1

Fix $x_0 \in G \setminus \bigcup_j \{a_j\}$. Choose $R > 0$ such that $\overline{B(x_0, 2R)} \subset G \setminus \bigcup_j \{a_j\}$. By Fubini's theorem we may find some $R' \in (R, 2R)$ such that (by passing to a further subsequence if necessary)

(13)
$$\int_{\partial B(x_0, R')} |\nabla u_{\varepsilon_n}|^2 \leq C$$

and

(14)
$$\int_{\partial B(x_0, R')} (|u_{\varepsilon_n}|^2 - 1)^2 \leq C\varepsilon_n^2.$$

From (13) we deduce that

$$u_{\varepsilon_n} \to u_\star \quad \text{uniformly on } \partial B(x_0, R').$$

Since

$$\deg\Big(u_{\varepsilon_n}, \partial B(x_0, R')\Big) = 0,$$

because $|u_{\varepsilon_n}| \geq 1/2$ in $B(x_0, R')$, we deduce that

(15)
$$\deg\Big(u_\star,\ \partial B(x_0, R')\Big) = 0.$$

We are now in a position to apply Theorem 2 in F. Bethuel, H. Brezis and F. Hélein [2] (which was written for this purpose!); see also Appendix I at the end of the book. This yields (5),(6),(10) and (11).

Property (9) follows from the fact that $u_\star$ is smooth away from the singularities.

In order to prove (8) it suffices, in view of Lemma V.2, to check that

(16)
$$\kappa_j = \deg(u_\star, a_j).$$

This is clear if $a_j \in G$, by (10). In case $a_j \in \partial G$ we may choose, as above, $R' > 0$ such that

$$\int_{\partial B(a_j, R')} |\nabla u_{\varepsilon_n}|^2 \leq C$$

and such that $\overline{B(a_j, R')}$ contains no other singularity. As above,

$$u_{\varepsilon_n} \to u_\star \quad \text{uniformly on } \partial B(a_j, R')$$

and thus

$$\kappa_j = \deg(u_{\varepsilon_n}, \partial B(a_j, R')) = \deg(u_\star, \partial B(a_j, R'))$$

for n large.

Proof of (12).

Step 1: $u_{\varepsilon_n} \to u_\star$ **in** $H^1_{\text{loc}}(G' \setminus \bigcup_j \{a_j\})$ **and in** $C^0_{\text{loc}}(G' \setminus \bigcup_j \{a_j\})$.

It suffices, in view of (10), to consider a point $x_0 \in \partial G$ that is not a singularity (a_j) and to show that for some R', u_{ε_n} converges to $u_\star$ in $H^1(B(x_0, R') \cap G)$ and in $C^0(\overline{B(x_0, R') \cap G})$.

Fix $R < \dfrac{1}{2}$ dist $(G, \partial G')$ such that $B(x_0, 2R)$ does not contain any singularity. By Fubini we may find $R' \in (R, 2R)$ such that (up to a subsequence)

$$\int_{\partial B(x_0, R')} |\nabla u_{\varepsilon_n}|^2 \leq C$$

and

$$\int_{\partial B(x_0,R')} (|u_{\varepsilon_n}|^2 - 1)^2 \le C\varepsilon_n^2.$$

Since $\deg(u_{\varepsilon_n}, \partial B(x_0, R')) = 0$ we may apply Theorem 2 of F. Bethuel, H. Brezis and F. Hélein [2] in $G \cap B(x_0, R')$ (see also Appendix I at the end of the book) to conclude that

$$u_{\varepsilon_n} \to u_\star \quad \text{in } H^1\Big(G \cap B(x_0, R')\Big) \cap C^0\Big(\overline{G \cap B(x_0, R')}\Big)$$

and also that

$$\frac{1}{\varepsilon_n^2} \int_{B(x_0,R')} (|u_{\varepsilon_n}|^2 - 1)^2 \to 0.$$

Step 2: $u_{\varepsilon_n} \to u_\star \quad$ in $C^{1,\alpha}_{\mathrm{loc}}(\bar{G} \setminus \bigcup_j \{a_j\})$.

We shall use the fact that the points a_j are not on the boundary ∂G. This will be proved in the next Section (and of course the argument there does not depend on Step 2). Let $U = G \setminus \bigcup B(a_j, \delta)$ with $\delta > 0$ sufficiently small. We already know that $u_{\varepsilon_n} \to u_\star$ in $H^1(U)$ and in $C(\bar{U})$. Following the same argument as in F. Bethuel, H. Brezis and F. Hélein [2] (part B in Section 2) we prove that u_ε is bounded in $H^2(U)$. As in Step B.4 of the above reference we let

$$\psi = \frac{1}{\varepsilon^2}(1 - |u_\varepsilon|^2)$$

and we have

$$-2\varepsilon^2 \Delta\psi + \psi \le 4|\nabla u_\varepsilon|^2 \qquad \text{in } U.$$

Multiplying this inequality by ψ^{q-1} we see that

$$\int_U \psi^q \le 4 \int_U |\nabla u_\varepsilon|^2 \psi^{q-1} + 2\varepsilon^2 \int_{\partial U} \frac{\partial \psi}{\partial \nu} \psi^{q-1}.$$

We split ∂U as $\partial U = \Gamma \cup \partial G$. Note that

$$\psi = 0 \quad \text{on } \partial G$$

while

$$\varepsilon^2 \Big|\frac{\partial \psi}{\partial \nu}\Big| \psi^{q-1} \le C \quad \text{on } \Gamma$$

since ψ is bounded in $L^\infty(\Gamma)$ by (11) and $\varepsilon^2 \dfrac{\partial \psi}{\partial \nu} = -\dfrac{\partial}{\partial \nu}|u|^2$ is bounded in $L^\infty(\Gamma)$ by (10). We then argue as in Step B.4 to conclude that ψ is bounded in $L^\infty(U)$.

VI.2. Further properties of $u_\star$: singularities have degree one and they are not on the boundary

The main result of this section is

Theorem VI.2. *We have*

$$\kappa_j = \deg(u_\star, a_j) = +1 \quad \forall j \tag{17}$$

$$a_j \in G \quad \forall j. \tag{18}$$

Consequently, there are **exactly** *d distinct points in the collection* (a_j).

Step 1: $\kappa_j = \deg(u_\star, a_j) > 0$.

We already know (see Lemma V.2) that $\kappa_j \geq 0, \quad \forall j$. We are going to prove that $\kappa_j = 0$ is impossible. Suppose not and say that for some j, $\kappa_j = 0$. We may find, as above, some R such that (by passing to a subsequence if necessary)

$$\int_{\partial B(a_j,R)} |\nabla u_{\varepsilon_n}|^2 \leq C,$$

$$\int_{\partial B(a_j,R)} (|u_{\varepsilon_n}|^2 - 1)^2 \leq C\varepsilon_n^2,$$

and moreover $\overline{B(a_j,R)}$ does not contain any other singularity, so that $\deg(u_{\varepsilon_n}, \partial B(a_j,R)) = \kappa_j = 0$. We may now apply Theorem 2, Step 1, of F. Bethuel, H. Brezis and F. Hélein [2] in $B(a_j,R) \cap G$ to conclude that

$$\frac{1}{\varepsilon_n^2}\int_{B(a_j,R)} (|u_{\varepsilon_n}|^2 - 1)^2 \to 0. \tag{19}$$

On the other hand, by the definition of a_j, there exists at least one bad disc $B(x_i, 2\lambda\varepsilon_n)$ contained in $B(a_j,R)$. Recall that (see the beginning of Section IV.1)

$$\frac{1}{\varepsilon_n^2}\int_{B(x_i,2\lambda\varepsilon_n)} (|u_{\varepsilon_n}|^2 - 1)^2 \geq \mu_0 > 0, \quad \text{for every } i. \tag{20}$$

Combining (19) and (20) we have a contradiction.

Step 2: $\kappa_j = 1$.

Fix $\eta > 0$, such that

$$\eta \leq \frac{1}{8} \operatorname*{Min}_{j \neq k} |a_j - a_k| + \frac{1}{2}\operatorname{dist}(G, \partial G').$$

Applying Corollary II.2 in G' with

$$\Omega = G' \setminus \bigcup_j B(a_j, \eta)$$

and $\mu = \frac{1}{2}\,\mathrm{dist}(G, \partial G')$, we obtain

$$\int_\Omega |\nabla u_\star|^2 \geq 2\pi \sum_j \kappa_j^2 \left(\log \frac{\mu}{\eta}\right) - C \tag{21}$$

where C depends only on d, G and G'. We may rewrite (21) as

$$\frac{1}{2}\int_\Omega |\nabla u_\star|^2 \geq \pi \left(\sum_j \kappa_j^2\right) |\log \eta| - C. \tag{22}$$

On the other hand, if we pass to the limit in Theorem V.1 as $n \to \infty$, we are led to

$$\frac{1}{2}\int_\Omega |\nabla u_\star|^2 \leq \pi\, d |\log \eta| + C, \tag{23}$$

where C in (22) and (23) is independent of η. Note that the passage to the limit is justified by Step 1 in the proof of Theorem VI.1.

Combining (22) and (23), we find

$$\sum_j (\kappa_j^2 - \kappa_j) |\log \eta| \leq C.$$

Letting $\eta \to 0$, we see that

$$\sum_j (\kappa_j^2 - \kappa_j) \leq 0.$$

This yields $\kappa_j = 1$ (by Step 1).

Step 3: $a_j \in G$, $\forall j$.

Up to now, the only information that we have is $a_j \in \bar{G}$. We are going to exclude the possibility that $a_j \in \partial G$. Assume not and say that $a_j \in \partial G$, for some j. For convenience suppose $a_1 \in \partial G$.

Let $R > 0$ be such that

$$\overline{B(a_1, R)} \subset G' \setminus \{a_2, \dots, a_d\}.$$

In what follows we choose $\eta \in (0, R)$, which will tend to zero, as in Step 2. Our next lemma plays an important role.

Lemma VI.1. *Let $a \in \partial G$. For every map u that belongs to $C^1_{\mathrm{loc}}\left(\overline{B(a,R)} \setminus \{a\}\,; S^1\right)$, such that*

$$u = \bar{g} \quad \text{in } (G' \setminus G) \cap B(a,R) \tag{24}$$

and

$$\deg\Big(u\,,\ \partial B(a,R)\Big) = 1 \tag{25}$$

we have

$$\frac{1}{2}\int_{B(a,R)\setminus B(a,\eta)} |\nabla u|^2 \geq 2\pi |\log \eta| - C, \quad \forall \eta \in (0,R)$$

where C depends only on $\bar{g}$ and R.

We postpone the proof of Lemma VI.1.

Proof of Step 3 completed. As in Step 2 we apply Corollary II.2 in $G' \setminus B(a_1,R)$ with $\Omega' = (G' \setminus B(a_1,R)) \setminus \bigcup_{j=2}^{d} B(a_j,\eta)$. We obtain

$$\frac{1}{2}\int_{\Omega'} |\nabla u_\star|^2 \geq \pi(d-1)|\log \eta| - C \tag{26}$$

where C depends only on R, d, G and G'. By Lemma VI.1 we have

$$\frac{1}{2}\int_{B(a_1,R)\setminus B(a_1,\eta)} |\nabla u_\star|^2 \geq 2\pi |\log \eta| - C. \tag{27}$$

Hence, combining (26) and (27), we see that

$$\frac{1}{2}\int_{\Omega} |\nabla u_\star|^2 \geq \pi(d+1)|\log \eta| - C$$

where $\Omega = G' \setminus \bigcup_{j=1}^{d} B(a_j,\eta)$, and C depends only on R, d, G, G' and R.

We now proceed as in the proof of Step 2, and conclude that

$$|\log \eta| \leq C, \quad \forall \eta \in (0,R),$$

where C is independent of η. Impossible.

Proof of Lemma VI.1. By a conformal change of variables we may always assume that, locally, G is the half-space $\{(x_1,x_2);\ x_2 > 0\}$, and that $a = (0,0)$.

In the conformal transformation $B(a,R)\setminus B(a,\eta)$ is transformed into a domain containing $B(0,R')\setminus B(0,\eta')$, with $R'\simeq R$ and $\eta'\simeq\eta$.

Consider a circle S_t of radius t centered at zero, with $\eta<t<R$.

We have

$$1=\deg(u,S_t)=\frac{1}{2\pi}\int_{S_t}u\times u_\tau=\frac{1}{2\pi}\int_{S_t^+}u\times u_\tau+\frac{1}{2\pi}\int_{S_t^-}\bar g\times\bar g_\tau$$

where $S_t^+=S_t\cap\{x_2>0\}$ and $S_t^-=S_t\cap\{x_2<0\}$. Note that

$$\left|\int_{S_t^-}\bar g\times\bar g_\tau\right|\le C|g(t,0)-g(-t,0)|\le C\sqrt{t},$$

since $g\in H^1(\partial G)$. Hence,

$$1-C\sqrt{t}\le\frac{1}{2\pi}\int_{S_t^+}|u_\tau|\le\frac{1}{2\pi}\left(\int_{S_t^+}|\nabla u|^2\right)^{1/2}(\pi t)^{1/2}.$$

Therefore

$$\frac{4\pi}{t}-\frac{C}{\sqrt{t}}\le\int_{S_t}|\nabla u|^2.$$

Integrating this inequality on (η,R) we obtain the desired conclusion.

CHAPTER VII

$u_\star$ coincides with THE canonical harmonic map having singularities (a_j)

In Section I.3 we have introduced the notion of a canonical harmonic map associated to given singularities with prescribed degrees. We shall only consider the case of prescribed degrees +1. We recall its main properties. Given d points $a_1, a_2, \ldots, a_d$ in G, let $\mathcal{C}$ be the class of all smooth harmonic maps from $G \setminus \{a_1, a_2, \ldots, a_d\}$ into S^1 such that

(i) $\deg(u, a_j) = +1 \quad \forall j \in \{1, 2, \ldots, d\}$

(ii) u is continuous up to ∂G and $u = g$ on ∂G.

Given any distinct points $a_1, a_2, \ldots, a_d$ in G, there exists a unique u_0 in $\mathcal{C} \cap W^{1,1}(G)$ satisfying in addition

$$\frac{\partial}{\partial x_1}\left(u_0 \times \frac{\partial u_0}{\partial x_1}\right) + \frac{\partial}{\partial x_2}\left(u_0 \times \frac{\partial u_0}{\partial x_2}\right) = 0 \text{ in } \mathcal{D}'(G). \tag{1}$$

This u_0 is called the canonical harmonic map associated to the prescribed singularities (a_j). This u_0 also satisfies

$$u_0(z) = \frac{(z - a_j)}{|z - a_j|} e^{i\psi_j(z)}, \text{ near } a_j \tag{2}$$

where ψ_j is a real valued harmonic function, smooth in some neighborhood of a_j (including a_j). In other words, $u_0(z)$ behaves like $e^{i(\theta+\theta_j)}$ for some constant phase $\theta_j = \psi_j(a_j)$.

The main result in this Chapter is the following:

Theorem VII.1. *Let a_j and $u_\star$ be as in Chapter VI, then $u_\star = u_0 =$ the canonical harmonic map associated to the singularities (a_j).*

Proof.

The strategy is the following: we first prove that $u_\star$ is in $W^{1,1}(G)$. By Remark I.1, we then know that

$$u_\star = u_0 \exp\left(i \sum_j c_j \log|x - a_j|\right) e^{i\chi}, \tag{3}$$

for some real constants c_j, and χ is the solution of

$$(4)\qquad \begin{cases} \Delta\chi = 0 & \text{in } G \\ \chi(x) + \sum_j c_j \log|x - a_j| = 0 & \text{on } \partial G. \end{cases}$$

The next step will be to prove that $c_j = 0,\ \forall j$. Hence $\chi = 0$, and $u_\star = u_0$.

Step 1: Consider for every n

$$\omega_n = \left|\frac{\partial u_{\varepsilon_n}}{\partial x_1}\right|^2 - \left|\frac{\partial u_{\varepsilon_n}}{\partial x_2}\right|^2 - 2i\frac{\partial u_{\varepsilon_n}}{\partial x_1}\cdot\frac{\partial u_{\varepsilon_n}}{\partial x_1},$$

which is the so-called Hopf differential. (Warning: the dot product refers to the scalar product of vectors, not the multiplication of complex numbers.) This quantity was first introduced in the framework of minimal surfaces, and plays a crucial role in many two-dimensional problems that are invariant under conformal transformations (see e.g., J. Sacks and K. Uhlenbeck [1], M. Grüter [1], R. Schoen [1]). Set

$$W_n = \frac{1}{4\varepsilon_n^2}(|u_{\varepsilon_n}|^2 - 1)^2.$$

A straightforward computation shows that any solution of

$$-\Delta u_{\varepsilon_n} = \frac{1}{\varepsilon_n^2} u_{\varepsilon_n}(1 - |u_{\varepsilon_n}|^2),$$

satisfies

$$(5)\qquad \frac{\partial \omega_n}{\partial \bar z} = \frac{\partial}{\partial z}(2W_n)$$

where, as usual,

$$\frac{\partial}{\partial \bar z} = \frac{1}{2}\left(\frac{\partial}{\partial x_1} + i\frac{\partial}{\partial x_2}\right) \quad\text{and}\quad \frac{\partial}{\partial z} = \frac{1}{2}\left(\frac{\partial}{\partial x_1} - i\frac{\partial}{\partial x_2}\right).$$

Lemma VII.1. *The sequence (W_n) is bounded in $L^1(G)$, and converges, up to a subsequence, in the weak $\star$ topology of $C(\bar G)$ to*

$$(6)\qquad W_\star = \sum_j m_j \delta_{a_j},\ \textit{with } m_j \ge 0.$$

In other words,

$$\int_G W_n(x)\zeta(x) \to \sum_j m_j \zeta(a_j),\quad \forall \zeta \in C(\bar G).$$

Proof of Lemma VII.1. The fact that W_n is bounded in $L^1(G)$ follows from Theorem III.2. Recall that, by (11) in Theorem VI.1,

$$(1-|u_{\varepsilon_n}|^2) \le C_K\,\varepsilon_n^2 \quad \text{in } K,$$

for every compact subset K in $G\setminus\bigcup_j\{a_j\}$. Therefore,

$$W_n \le C_K^2\varepsilon_n^2 \quad \text{in } K,$$

and consequently W_n converges, up to a subsequence, to a measure μ, with

$$\text{supp }(\mu) \subset \partial G \cup \Big(\bigcup_j\{a_j\}\Big).$$

On the other hand, we know (see Step 1 in the proof of Theorem VI.1) that for every point x_0 on ∂G, there exists a constant $R' > 0$, such that

$$\int_{B(x_0,R')} W_n \to 0;$$

this implies that $\text{supp}(\mu) \subset \bigcup_j\{a_j\}$, and μ has the required form.

Back to Step 1: Consider the distribution

$$T = \frac{\partial}{\partial z}\left(\frac{1}{\pi z}\right) = \frac{\partial^2}{\partial z^2}\left(\frac{2}{\pi}\log r\right).$$

Set

$$\alpha_n = T * \overline{W}_n \qquad \text{in the sense of distributions,}$$

where $\overline{W}_n = W_n$ in G, and $\overline{W}_n = 0$ outside G.

We have

$$\frac{\partial \alpha_n}{\partial \bar z} = \frac{\partial \overline{W}_n}{\partial z} \quad \text{in } \mathcal{D}'(\mathbf{R}^2), \tag{7}$$

since $\dfrac{\partial}{\partial \bar z}\left(\dfrac{1}{\pi z}\right) = \dfrac{\partial^2}{\partial \bar z\partial z}\left(\dfrac{2}{\pi}\log r\right) = \dfrac{1}{2\pi}\Delta(\log r) = \delta_0.$

From (5) and (7) we obtain that

$$\frac{\partial}{\partial \bar z}(\omega_n - 2\alpha_n) = 0 \quad \text{in } \mathcal{D}'(G). \tag{8}$$

Set

$$\beta_n = \omega_n - 2\alpha_n.$$

Let K be a compact subset in $G \setminus \bigcup_j \{a_j\}$.

Claim: β_n is bounded in $L^\infty(K)$.

Proof of the Claim. Actually, both ω_n and α_n are bounded. The boundedness of ω_n follows from the boundedness of u_{ε_n} in $C^1(K)$ (see Theorem VI.1).

To prove the boundedness of α_n, set

$$N_R(K) = \{x \in G;\ \ \text{dist}(x, K) \le R\}$$

and choose R sufficiently small in order to have

$$\overline{W}_n \quad \text{bounded in} \quad C^2(N_R(K))$$

(this is possible by (11) in Theorem VI.1). Fix a function $\zeta \in C_c^\infty(B(0,R))$ such that $\zeta \equiv 1$ in $B(0, R/2)$ and write

$$\alpha_n = (\zeta T) * \overline{W}_n + [(1-\zeta)T] * \overline{W}_n.$$

Clearly $(\zeta T) * \overline{W}_n$ is bounded in $L^\infty(K)$. On the other hand, $[(1-\zeta)T] * \overline{W}_n$ is bounded in $L^\infty(\mathbb{R}^2)$ since $(1-\zeta)T \in L^\infty(\mathbb{R}^2)$ and $\overline{W}_n$ is bounded in $L^1(\mathbb{R}^2)$. This completes the proof of the Claim.

Since β_n is a holomorphic function on G, it is bounded in $C^k_{\text{loc}}(G)$, $\forall k$. Thus, up to a subsequence, we may assume that

$$\beta_n \to \beta \quad \text{in } C^k_{\text{loc}}(G), \quad \forall k, \tag{9}$$

for some holomorphic function β on G.

On the other hand, since $\overline{W}_n$ converges to $W_\star$ in $\mathcal{D}'(\mathbb{R}^2)$, we conclude that

$$\alpha_n \to \alpha_\star = T * W_\star \quad \text{in } \mathcal{D}'(\mathbb{R}^2). \tag{10}$$

Thus,

$$\omega_n = \beta_n + 2\alpha_n \to \beta + 2\alpha_\star \quad \text{in } \mathcal{D}'(G). \tag{11}$$

From the definition of ω_n, it is clear that

$$\omega_n \to \omega_\star \equiv \left|\frac{\partial u_\star}{\partial x_1}\right|^2 - \left|\frac{\partial u_\star}{\partial x_2}\right|^2 - 2i\frac{\partial u_\star}{\partial x_1}\cdot\frac{\partial u_\star}{\partial x_2} \tag{12}$$

and this convergence holds in $C^k_{\rm loc}(G \setminus \bigcup_j \{a_j\})$.

From (11) and (12) we obtain

$$\omega_\star = \beta + 2\alpha_\star, \quad \text{in } \mathcal{D}'(G \setminus \bigcup_j \{a_j\}). \tag{13}$$

We now derive further properties of $\omega_\star$ and $\alpha_\star$. Because of (6) we have in $\mathcal{D}'(G \setminus \bigcup_j \{a_j\})$

$$\alpha_\star = \sum_j m_j T * \delta_{a_j} = -\sum_j \frac{m_j}{\pi(z-a_j)^2}. \tag{14}$$

On the other hand, we have

$$|\omega_\star| = |\nabla u_\star|^2 \quad \text{in } G \setminus \bigcup_j \{a_j\}. \tag{15}$$

Indeed, by (12),

$$|\omega_\star|^2 = \left(\left| \frac{\partial u_\star}{\partial x_1} \right|^2 - \left| \frac{\partial u_\star}{\partial x_2} \right|^2 \right)^2 + 4 \left(\frac{\partial u_\star}{\partial x_1} \cdot \frac{\partial u_\star}{\partial x_2} \right)^2,$$

and (15) follows.

Applying (13), (14) and (15), we find

$$|\nabla u_\star|^2 \le |\beta| + \sum_j \frac{2m_j}{\pi |z-a_j|^2}. \tag{16}$$

An important consequence of (16) is that

$$|\nabla u_\star| \le \frac{C}{|z-a_j|} \quad \text{near each } a_j; \tag{17}$$

in particular,

$$u_\star \in W^{1,p}(G), \quad \forall p < 2. \tag{18}$$

Step 2: $u_\star = u_0$.

We are now in position to apply Remark I.1, which asserts that (3) holds, for some real constants c_j. We will now prove that

$$c_j = 0, \quad \forall j.$$

Fix one of the points a_j; for simplicity take $a_j = 0$. Recall that, in polar coordinates, (see (34) in Chapter I),

(19) $$u_0 = e^{i\theta}e^{iS}, \quad \text{near zero},$$

where S is a harmonic function. From (3) and (19) it follows that

(20) $$u_\star = \exp[i\theta + i\, c_j \log\, r + i\, \chi'], \quad \text{near zero},$$

where χ' is some harmonic function near zero (and including zero). We now compute $\omega_\star$ using (20). Note that if, locally, $u = e^{i\varphi}$ then

$$\omega = \left|\frac{\partial u}{\partial x_1}\right|^2 - \left|\frac{\partial u}{\partial x_2}\right|^2 - 2i\frac{\partial u}{\partial x_1}\cdot\frac{\partial u}{\partial x_2} = \left(\frac{\partial \varphi}{\partial x_1} - i\frac{\partial \varphi}{\partial x_2}\right)^2.$$

Applying this with $\varphi = \theta + c_j \log\, r + \chi'$, we obtain

(21) $$\omega_\star = \left[(c_j - i)\frac{1}{z} + \chi''\right]^2,$$

where $\chi'' = 2\dfrac{\partial \chi'}{\partial z}$. We expand (21) and compare with (13) and (14); this yields, for z near 0, $z \neq 0$,

(22) $$\frac{(c_j - i)^2}{z^2} + \frac{2\chi''(c_j - i)}{z} + (\chi'')^2 = -\frac{2m_j}{\pi z^2} + \beta',$$

with $\beta' = \beta - \sum_{\ell \neq j} \dfrac{2m_\ell}{\pi(z - a_\ell)^2}$.

Multiplying (22) by z^2 and letting $z \to 0$, we deduce that

(23) $$(c_j - i)^2 = -\frac{2m_j}{\pi}.$$

Since c_j is real we see that

(24) $$c_j = 0 \quad \text{and} \quad m_j = \frac{\pi}{2}.$$

During the course of this proof, we have also established the following theorem:

Theorem VII.2. *We have, in the weak $\star$ topology of $C(\bar{G})$,*

(25) $$\frac{1}{4\varepsilon_n^2}\left(|u_{\varepsilon_n}|^2 - 1\right)^2 \to \frac{\pi}{2}\sum_j \delta_{a_j}.$$

In the same spirit, the following holds:

Theorem VII.3. *We have, in the weak $\star$ topology of $C(\overline{G})$,*

$$\frac{1}{|\log \varepsilon_n|}|\nabla u_{\varepsilon_n}|^2 \to 2\pi \sum_j \delta_{a_j}. \tag{26}$$

Proof. In view of Theorem III.1 we know that

$$\int_G |\nabla u_\varepsilon|^2 \le 2\pi d\, |\log \varepsilon| + C. \tag{27}$$

On the other hand, Theorem VI.1 tells us that

$$\frac{1}{|\log \varepsilon_n|}|\nabla u_{\varepsilon_n}|^2 \to 0 \quad \text{in } L^1(K)$$

for every compact subset $K \subset \bar{G} \setminus \bigcup\{a_j\}$.
Consequently (up to a subsequence)

$$\frac{1}{|\log \varepsilon_n|}|\nabla u_{\varepsilon_n}|^2 \to \sum_j m_j \delta_{a_j} \tag{28}$$

with

$$\sum_j m_j \le 2\pi d \quad \text{by (27)}. \tag{29}$$

From Theorem V.2 and Theorem VI.2 we deduce that, for every j and every $\eta > 0$,

$$\int_{B(a_j,\eta)} |\nabla u_{\varepsilon_n}|^2 \ge 2\pi\, |\log \varepsilon_n| - C(\eta) \quad \forall n \ge N(\eta). \tag{30}$$

It follows from (30) that

$$m_j \ge 2\pi \qquad \forall j. \tag{31}$$

Combining (31) with (29) we conclude that $m_j = 2\pi$, $\forall j$ and the desired result follows.

The last result of this chapter is

Theorem VII.4. *We have, near each singularity a_j,*

$$u_\star(z) = \frac{z - a_j}{|z - a_j|} e^{i\,H_j(z)}, \tag{32}$$

where H_j is a real harmonic function such that

$$H_j(z) = H_j(a_j) + O(|z - a_j|^2), \text{ as } z \to a_j.$$

In other words,

$$\nabla H_j(a_j) = 0. \tag{33}$$

Remark VII.1. Note that property (33) does not hold for the canonical harmonic map u_0 associated to an **arbitrary** configuration of points (a_j). This additional property is related to the fact that $u_\star$ is the limit of a sequence of minimizers for E_ε, and that the configuration (a_j) minimizes the renormalized energy W (see Section VIII.2).

Proof of Theorem VII.4. In view of (22) and (23) we must have, assuming $a_j = 0$,

$$\chi''(0) = 2\frac{\partial \chi'}{\partial z}(0) = 0$$

(here we have used the fact that $c_j = 0$ and $m_j = \pi/2$).

This means, since χ' is real valued, that

$$\nabla \chi'(0) = 0.$$

Recall that, near zero, $u_\star = e^{i\varphi}$ with $\varphi = \theta + \chi'$ and the desired conclusion follows.

Remark VII.2. The method used in the proof of Theorem VII.1 is in the same spirit as the argument of H. Brezis and L. Peletier [1] concerning the equation $-\Delta u = u^p$ where p is the critical exponent $(N+2)/(N-2)$. There, precise information about the blow-up behavior—in particular the location of the blow-up points—is obtained by multiplying the equation by $\dfrac{\partial u}{\partial x_i}$ and using various Pohozaev-type identities.

We now present alternative proofs of Theorems VII.2 and VII.4 assuming Theorems VI.1, VI.2 and VII.1. Recall that u_ε satisfies

$$-\Delta u_\varepsilon = \frac{1}{\varepsilon^2} u_\varepsilon (1 - |u_\varepsilon|^2) \quad \text{in } G. \tag{34}$$

1. Alternative proof of Theorem VII.2. Fix one of the points a_i and set $B_R = B(a_i, R)$ with R sufficiently small, so that B_R does not contain any of the other points a_j, $j \neq i$. For simplicity, assume that $a_i = 0$. Multiplying (34) by $\sum_{k=1}^{2} x_k \frac{\partial u_\varepsilon}{\partial x_k}$ and integrating on B_R we find (as in Section III.3)

$$(35)\qquad \begin{cases} \dfrac{R}{2}\displaystyle\int_{\partial B_R}\left|\frac{\partial u_\varepsilon}{\partial \nu}\right|^2 + \frac{1}{2\varepsilon^2}\int_{B_R}\left(|u_\varepsilon|^2-1\right)^2 = \\ \dfrac{R}{2}\displaystyle\int_{\partial B_R}\left|\frac{\partial u_\varepsilon}{\partial \tau}\right|^2 + \frac{R}{4\varepsilon^2}\int_{\partial B_R}\left(|u_\varepsilon|^2-1\right)^2. \end{cases}$$

Passing to the limit in (35) (using (10) of Theorem VI.1) yields

$$(36)\qquad \frac{R}{2}\int_{\partial B_R}\left|\frac{\partial u_\star}{\partial \nu}\right|^2 + 2m_i = \frac{R}{2}\int_{\partial B_R}\left|\frac{\partial u_\star}{\partial \tau}\right|^2 .$$

(Recall that $\frac{1}{4\varepsilon_n^2}(|u_{\varepsilon_n}|^2-1)^2$ converges to $\sum_j m_j \delta_{a_j}$.)

Since $u_\star$ is the canonical harmonic map associated to (a_j) and $\deg(u_\star, a_j) = +1\ \ \forall j$ we know that

$$(37)\qquad u_\star = e^{i(\theta+H)} \quad \text{in } B_R$$

where H is a smooth real-valued harmonic function in B_R. Thus, we have

$$(38)\qquad \left|\frac{\partial u_\star}{\partial \nu}\right|^2 = \left|\frac{\partial \theta}{\partial \nu} + \frac{\partial H}{\partial \nu}\right|^2 = \left|\frac{\partial H}{\partial \nu}\right|^2$$

and

$$(39)\qquad \left|\frac{\partial u_\star}{\partial \tau}\right|^2 = \left|\frac{\partial \theta}{\partial \tau} + \frac{\partial H}{\partial \tau}\right|^2 = \left|\frac{1}{R} + \frac{\partial H}{\partial \tau}\right|^2 = \frac{1}{R^2} + \frac{2}{R}\frac{\partial H}{\partial \tau} + \left|\frac{\partial H}{\partial \tau}\right|^2 .$$

Inserting (38) and (39) into (36) we find

$$(40)\qquad \frac{R}{2}\int_{\partial B_R}\left|\frac{\partial H}{\partial \nu}\right|^2 + 2m_i = \pi + \frac{R}{2}\int_{\partial B_R}\left|\frac{\partial H}{\partial \tau}\right|^2 .$$

On the other hand, if we multiply the equation

$$(41)\qquad \Delta H = 0$$

by $\sum_{k=1}^{2} x_k \dfrac{\partial H}{\partial x_k}$ and integrate on B_R we obtain

$$\frac{R}{2}\int_{\partial B_R}\left|\frac{\partial H}{\partial \nu}\right|^2 = \frac{R}{2}\int_{\partial B_R}\left|\frac{\partial H}{\partial \tau}\right|^2 . \tag{42}$$

Comparing (40) and (42) yields $m_i = \pi/2$ which is the desired conclusion.

2. Alternative proof of Theorem VII.4. Multiplying (34) by $\dfrac{\partial u_\varepsilon}{\partial x_k}$ and integrating we find

$$\begin{aligned}-\int_{\partial B_R}\frac{\partial u_\varepsilon}{\partial \nu}\cdot\frac{\partial u_\varepsilon}{\partial x_k}+\frac{1}{2}\int_{\partial B_R}|\nabla u_\varepsilon|^2(\nu\cdot e_k)\\ = -\frac{1}{4\varepsilon^2}\int_{\partial B_R}(|u_\varepsilon|^2-1)^2(\nu\cdot e_k).\end{aligned} \tag{43}$$

Passing to the limit in (43) (using (10) and (11) of Theorem VI.1) yields

$$-\int_{\partial B_R}\frac{\partial u_\star}{\partial \nu}\cdot\frac{\partial u_\star}{\partial x_k}+\frac{1}{2}\int_{\partial B_R}|\nabla u_\star|^2(\nu\cdot e_k)=0. \tag{44}$$

But we have

$$\left\{\begin{aligned}|\nabla u_\star|^2 &= |\nabla(\theta+H)|^2 = |\frac{\partial}{\partial \nu}(\theta+H)|^2+|\frac{\partial}{\partial \tau}(\theta+H)|^2\\ &= |\frac{\partial H}{\partial \nu}|^2+|\frac{1}{R}+\frac{\partial H}{\partial \tau}|^2=\frac{1}{R^2}+\frac{2}{R}\frac{\partial H}{\partial \tau}+|\nabla H|^2\end{aligned}\right. \tag{45}$$

and

$$\frac{\partial u_\star}{\partial \nu}\cdot\frac{\partial u_\star}{\partial x_k}=\frac{\partial}{\partial \nu}(\theta+H)\frac{\partial}{\partial x_k}(\theta+H)=\frac{\partial H}{\partial \nu}\Big(\frac{\tau\cdot e_k}{R}+\frac{\partial H}{\partial x_k}\Big). \tag{46}$$

Inserting (45) and (46) into (44) we obtain

$$-\int_{\partial B_R}\frac{\partial H}{\partial \nu}\left(\frac{\tau\cdot e_k}{R}+\frac{\partial H}{\partial x_k}\right)+\int_{\partial B_R}\left(\frac{1}{R}\frac{\partial H}{\partial \tau}+\frac{1}{2}|\nabla H|^2\right)(\nu\cdot e_k)=0. \tag{47}$$

On the other hand, if we multiply (41) by $\dfrac{\partial H}{\partial x_k}$ and integrate on B_R we are led to

$$-\int_{\partial B_R}\frac{\partial H}{\partial \nu}\frac{\partial H}{\partial x_k}+\frac{1}{2}\int_{\partial B_R}|\nabla H|^2(\nu\cdot e_k)=0. \tag{48}$$

Comparing (47) and (48) yields

$$-\int_{\partial B_R} \frac{\partial H}{\partial \nu}(\tau \cdot e_k) + \int_{\partial B_R} \frac{\partial H}{\partial \tau}(\nu \cdot e_k) = 0.$$

Since this is true for $k = 1, 2$ we find

$$-\int_{\partial B_R} \frac{\partial H}{\partial \nu}\tau + \int_{\partial B_R} \frac{\partial H}{\partial \tau}\ \nu = 0. \tag{49}$$

But

$$-\frac{\partial H}{\partial \nu}\tau + \frac{\partial H}{\partial \tau}\nu = (\nabla H)^\perp = \frac{\partial H}{\partial x_2}e_1 - \frac{\partial H}{\partial x_1}e_2. \tag{50}$$

From (49), (50) and the mean-value theorem (recall that $\frac{\partial H}{\partial x_1}$ and $\frac{\partial H}{\partial x_2}$ are harmonic functions) we deduce that

$$\nabla H(0) = 0$$

which is the desired conclusion.

CHAPTER VIII

The configuration (a_j) minimizes the renormalized energy W

In Section I.4 we have introduced the renormalized energy $W = W(a, d, g)$ for a general configuration of points (a_i), $1 \leq i \leq n$, with associated integers $d_i \in \mathbf{Z}$ such that $\sum_{i=1}^{n} d_i = d$. Throughout this chapter we consider only configurations consisting of **exactly d distinct points in G, each one associated to $d_i = +1$.**

VIII.1. The general case

The main result in this section is the following:

Theorem VIII.1. *Let (a_j) be as in Chapter VI, then the configuration (a_j) minimizes W.*

The proof relies on the following two lemmas:

Lemma VIII.1. *Let $\bar{a} = (\bar{a}_j)$ be any configuration of d distinct points in G. Then there is some $\rho_0 > 0$ (depending only on $\bar{a}$ and G) such that, for every $\rho < \rho_0$ and every $\varepsilon > 0$, we have*

$$E_\varepsilon(u_\varepsilon) \leq d\, I(\varepsilon, \rho) + W(\bar{a}) + \pi d \log(1/\rho) + O(\rho), \tag{1}$$

where $O(\rho)$ stands for a quantity X such that $|X| \leq C\rho$ and C depends only on G, $\bar{a}$ and g.

Lemma VIII.2. *Let $a = (a_j)$ be as in Chapter VI. Then, given any ρ (sufficiently small, say $\rho < \rho_1$), there is an integer $N = N(\rho)$ such that, for every $n \geq N$,*

$$E_{\varepsilon_n}(u_{\varepsilon_n}) \geq d\, I(\varepsilon_n, \rho) + W(a) + \pi d \log(1/\rho) + O(\rho^2), \tag{2}$$

where $O(\rho^2)$ stands for a quantity Y such that $|Y| \leq C\rho^2$ and C depends only on G, a and g.

Recall that $I(\varepsilon, \rho)$ has been defined in Section III.1.

We postpone the proof of the lemmas and present the proof of the theorem.

Proof of Theorem VIII.1. Fix $\rho < \min\{\rho_0, \rho_1\}$. Combining (1) and (2) we obtain

$$W(a) \le W(\bar{a}) + O(\rho). \tag{3}$$

Letting $\rho \to 0$ we see that

$$W(a) \le W(\bar{a})$$

and since $\bar{a}$ is an arbitrary configuration we conclude that a is a minimizing configuration for W.

Proof of Lemma VIII.1. We apply Theorem I.9 to the configuration $\bar{a}$. This yields, for every $\rho < \rho_0 = \min_{j \neq k} \left\{\frac{1}{2}|\bar{a}_j - \bar{a}_k|, \mathrm{dist}(\bar{a}_j, \partial G)\right\}$, some map $\hat{u}_\rho$ from Ω_ρ to S^1 such that $\hat{u}_\rho = g$ on ∂G, $\hat{u}_\rho(z) = \alpha_j \dfrac{(z - \bar{a}_j)}{|z - \bar{a}_j|}$ on $\partial B(\bar{a}_j, \rho)$ with $|\alpha_j| = 1$ and

$$\frac{1}{2}\int_{\Omega_\rho} |\nabla \hat{u}_\rho|^2 = \pi d \log(1/\rho) + W(\bar{a}) + O(\rho). \tag{4}$$

On the other hand, for each j, by definition of $I(\varepsilon, \rho)$ (see Section III.1), we may find some $v_j : B(a_j, \rho) \to \mathbb{C}$ such that $v_j(z) = \alpha_j \dfrac{(z - \bar{a}_j)}{|z - \bar{a}_j|}$ on $\partial B(\bar{a}_j, \rho)$ and

$$\frac{1}{2}\int_{B(\bar{a}_j,\rho)} |\nabla v_j|^2 + \frac{1}{4\varepsilon^2}\int_{B(\bar{a}_j,\rho)} (|v_j|^2 - 1)^2 = I(\varepsilon, \rho). \tag{5}$$

Set

$$w = \begin{cases} \hat{u}_\rho & \text{in } \Omega_\rho \\ v_j & \text{in } B(\bar{a}_j, \rho), \quad j = 1, 2, \ldots, d. \end{cases}$$

Combining (4) and (5) we see that

$$E_\varepsilon(w) = d\, I(\varepsilon, \rho) + W(\bar{a}) + \pi d \, \log(1/\rho) + O(\rho)$$

and the desired conclusion follows.

Proof of Lemma VIII.2. Recall that u_{ε_n} converges in $H^1_{\mathrm{loc}}(\bar{G} \setminus \bigcup_j \{a_j\})$ to $u_\star$ (see Theorem VI.1) and therefore, for every fixed ρ, $\rho < \rho_1 = \min_{j \neq k} \left\{\frac{1}{2}|a_j - a_k|, \mathrm{dist}(a_j, \partial G)\right\}$,

$$\frac{1}{2}\int_{\Omega_\rho} |\nabla u_{\varepsilon_n}|^2 \to \frac{1}{2}\int_{\Omega_\rho} |\nabla u_\star|^2. \tag{6}$$

In particular, there is an integer $N_1 = N_1(\rho)$, such that, for every $n \geq N_1$,

$$\frac{1}{2}\int_{\Omega_\rho} |\nabla u_{\varepsilon_n}|^2 \geq \frac{1}{2}\int_{\Omega_\rho} |\nabla u_*|^2 - \rho^2. \tag{7}$$

On the other hand, recall (see Theorems I.8, VI.2 and VII.1) that

$$\frac{1}{2}\int_{\Omega_\rho} |\nabla u_*|^2 = \pi d \log(1/\rho) + W(a) + O(\rho^2). \tag{8}$$

Combining (7) and (8) we see that, for $n \geq N_1(\rho)$, we have

$$\frac{1}{2}\int_{\Omega_\rho} |\nabla u_{\varepsilon_n}|^2 + \frac{1}{4\varepsilon_n^2}\int_{\Omega_\rho} (|u_{\varepsilon_n}|^2 - 1)^2 \geq \pi d \log(1/\rho) + W(a) + O(\rho^2). \tag{9}$$

We now turn to energy estimates on the balls $B(a_j, \rho)$. We claim that given any ρ, $\rho < \rho_1$, there is an integer $N_2 = N_2(\rho)$ such that, for every $n \geq N_2$,

$$\frac{1}{2}\int_{B(a_j,\rho)} |\nabla u_{\varepsilon_n}|^2 + \frac{1}{4\varepsilon_n^2}\int_{B(a_j,\rho)} (|u_{\varepsilon_n}|^2 - 1)^2 \geq I(\varepsilon_n, \rho) + O(\rho^2). \tag{10}$$

Combining (9) and (10) we are led to the conclusion of Lemma VIII.2. Therefore, it remains only to prove the Claim.

Proof of Claim (10). Recall (see Theorem VII.2) that near each a_j we have

$$u_*(z) = e^{i(\theta + H_j(z))} \tag{11}$$

where $e^{i\theta} = \dfrac{z - a_j}{|z - a_j|}$ and H_j is a real-valued smooth harmonic function in a neighborhood of a_j, including a_j, with

$$\nabla H_j(a_j) = 0. \tag{12}$$

By Theorem VI.1 we know that, given any $\rho < \rho_1$, we may find some integer $N_3 = N_3(\rho)$ such that, for every $n \geq N_3$, we have

$$\left\| u_{\varepsilon_n} - u_* \right\|_{L^\infty(B(a_j,\rho)\setminus B(a_j,\rho/2))} \leq \rho^2 \tag{13}$$

and

$$\|\nabla u_{\varepsilon_n} - \nabla u_\star\|_{L^\infty(B(a_j,\rho)\setminus B(a_j,\rho/2))} \le \rho. \tag{14}$$

Using the same theorem we may also assume that, for $n \ge N_3$, we have

$$\frac{1-|u_{\varepsilon_n}|^2}{\varepsilon_n^2} \le |\nabla u_\star|^2 + 1 \quad \text{in } B(a_j,\rho)\setminus B(a_j,\rho/2). \tag{15}$$

From (11) and (12) we see that

$$|\nabla u_\star| \le \frac{2}{\rho} + O(\rho) \quad \text{in } B(a_j,\rho)\setminus B(a_j,\rho/2). \tag{16}$$

Combining (15) and (16) we find, for every $n \ge N_3$,

$$\frac{1-|u_{\varepsilon_n}|^2}{\varepsilon_n^2} \le \frac{4}{\rho^2} + C \equiv K(\rho) \quad \text{in } B(a_j,\rho)\setminus B(a_j,\rho/2) \tag{17}$$

where C denotes—here and throughout the rest of this section—a constant **independent of n and ρ.**

Consider the map

$$w_n(z) = \left(\frac{2|z-a_j|}{\rho} - 1\right)\left(e^{i(\theta+H_j(a_j))} - u_{\varepsilon_n}(z)\right) + u_{\varepsilon_n}(z) \tag{18}$$

defined for $z \in B(a_j,\rho)\setminus B(a_j,\rho/2)$. We summarize its main properties in the following lemma:

Lemma VIII.3. *There is an integer $N_4 = N_4(\rho)$ such that, for every $n \ge N_4$,*

$$\|w_n - u_{\varepsilon_n}\|_{L^\infty(B(a_j,\rho)\setminus B(a_j,\rho/2))} \le C\rho^2, \tag{19}$$

$$\|\nabla w_n - \nabla u_{\varepsilon_n}\|_{L^\infty(B(a_j,\rho)\setminus B(a_j,\rho/2))} \le C\rho \tag{20}$$

and

$$|w_n(z)|^2 \ge 1 - C\rho^4 \quad \forall z \in B(a_j,\rho)\setminus B(a_j,\rho/2). \tag{21}$$

Proof of Lemma VIII.3. We have, by (18),

$$|w_n - u_{\varepsilon_n}| \le \left|e^{i(\theta+H_j(a_j))} - u_\star\right| + \left|u_\star - u_{\varepsilon_n}\right|.$$

Hence, we deduce (19) from (11), (12) and (13), for every $n \geq N_3$. Differentiating (18) we easily see that

$$\left|\nabla w_n - \nabla u_{\varepsilon_n}\right| \leq \frac{2}{\rho}\left|e^{i(\theta+H(a_j))} - u_{\varepsilon_n}\right| + \left|\nabla e^{i(\theta+H(a_j))} - \nabla u_{\varepsilon_n}\right|.$$

Applying (11), (12), (13) and (14) we are led to (20) for every $n \geq N_3$.

The proof of (21) relies on the following variant of the parallelogram identity

$$(22)\quad \begin{aligned} |t\,a + (1-t)b|^2 &= t|a|^2 + (1-t)|b|^2 - t(1-t)|a-b|^2 \\ &\geq t|a|^2 + (1-t)|b|^2 - \frac{1}{4}|a-b|^2, \quad \forall t \in [0,1]. \end{aligned}$$

We apply (22) with $a = e^{i(\theta+H_j(a_j))}$, $b = u_{\varepsilon_n}(z)$ and $t = \left(\frac{2|z-a_j|}{\rho} - 1\right)$; this yields, using (17),

$$(23)\qquad |w_n(z)|^2 \geq 1 - 2K(\rho)\varepsilon_n^2 - \frac{1}{4}|e^{i(\theta+H_j(a_j))} - u_{\varepsilon_n}(z)|^2.$$

We finally choose $N_4(\rho) \geq N_3(\rho)$ such that

$$(24)\qquad K(\rho)\varepsilon_n^2 \leq \rho^4 \quad \forall n \geq N_4(\rho)$$

and then (21) follows from (23), (24), (11), (12) and (13).

We may now return to the proof of Claim (10).

Proof of Claim (10) completed. Set

$$(25)\qquad R = R(n,\rho) = \sqrt{1 - K(\rho)\varepsilon_n^2}.$$

We may always assume that $n \geq N_4(\rho)$ and $\rho < 1$ so that R is well defined. Consider the map $P = P(n,\rho)$ from $\mathbb{C} \setminus \{0\}$ into itself defined by:

$$P\xi = \begin{cases} \xi & \text{if } |\xi| \geq R, \\ R\dfrac{\xi}{|\xi|} & \text{if } |\xi| < R. \end{cases}$$

A standard computation shows that

$$(26)\qquad \|\nabla P(\xi)\| \leq \begin{cases} 1 & \text{if } |\xi| \geq R, \\ \dfrac{R}{|\xi|} & \text{if } |\xi| < R. \end{cases}$$

Consider the map $v_n : B(a_j, \rho) \to \mathbb{C}$ defined by

$$(27) \qquad v_n(z) = \begin{cases} u_{\varepsilon_n}(z) & \text{if } z \in B(a_j, \rho/2), \\ Pw_n(z) & \text{if } z \in B(a_j, \rho) \setminus B(a_j, \rho/2). \end{cases}$$

On $\partial B(a_j, \rho)$ we have $w_n(z) = e^{i(\theta + H_j(a_j))}$ and thus

$$v_n(z) = e^{i(\theta + H_j(a_j))}.$$

It follows from the definition of $I(\varepsilon_n, \rho)$ that

$$(28) \qquad \frac{1}{2}\int_{B(a_j,\rho)} |\nabla v_n|^2 + \frac{1}{4\varepsilon_n^2}\int_{B(a_j,\rho)} (|v_n|^2 - 1)^2 \geq I(\varepsilon_n, \rho).$$

[Note that on $\partial B(a_j, \rho/2)$, $w_n = u_{\varepsilon_n}$ and thus $v_n = u_{\varepsilon_n}$ since $|u_{\varepsilon_n}| \geq R$ by (17); hence $v_n \in H^1(B(a_j, \rho))$].

From (27) and (28) we deduce that

$$(29) \quad \frac{1}{2}\int_{B(a_j,\rho)} |\nabla u_{\varepsilon_n}|^2 + \frac{1}{4\varepsilon_n^2}\int_{B(a_j,\rho)} (|u_{\varepsilon_n}|^2 - 1)^2 \geq I(\varepsilon_n, \rho) - U - V$$

where $U = U(n, \rho)$ and $V = V(n, \rho)$ are defined by

$$(30) \qquad U = \frac{1}{2}\int_{B(a_j,\rho)\setminus B(a_j,\rho/2)} |\nabla v_n|^2 - |\nabla u_{\varepsilon_n}|^2$$

and

$$(31) \qquad V = \frac{1}{4\varepsilon_n^2}\int_{B(a_j,\rho)\setminus B(a_j,\rho/2)} (|v_n|^2 - 1)^2 - (|u_{\varepsilon_n}|^2 - 1)^2.$$

We first estimate V. Since $|w_n| \leq 1$ (because w_n is a convex combination of u_{ε_n} and $e^{i(\theta + H_j(a_j))}$) we have $R \leq |v_n| \leq 1$ and therefore

$$(|v_n|^2 - 1)^2 \leq (1 - R^2)^2 = K^2(\rho)\varepsilon_n^4.$$

It follows from (17) and (24) that

$$(32) \qquad V \leq \pi\rho^2 K^2(\rho)\varepsilon_n^2 \leq C\rho^4 \quad \forall n \geq N_4(\rho).$$

We now estimate U. We have, in $B(a_j, \rho) \setminus B(a_j, \rho/2)$,

$$(33) \qquad |\nabla v_n|^2 \leq \|\nabla P(w_n)\|^2 |\nabla w_n|^2 \leq \frac{1}{(1 - C\rho^4)} |\nabla w_n|^2$$

by (26) and (21). From (33) and (20) we deduce that

$$(34)\qquad |\nabla v_n|^2 \le \frac{1}{(1-C\rho^4)}(|\nabla u_{\varepsilon_n}|^2 + 2C\rho|\nabla u_{\varepsilon_n}| + C^2\rho^2).$$

Hence

$$(35)\qquad |\nabla v_n|^2 - |\nabla u_{\varepsilon_n}|^2 \le C(\rho^4|\nabla u_{\varepsilon_n}|^2 + \rho|\nabla u_{\varepsilon_n}| + \rho^2).$$

On the other hand, it follows from (11) and (12) that

$$(36)\qquad \|\nabla u_\star\|_{L^\infty(B(a_j,\rho)\setminus B(a_j,\rho/2))} \le \frac{C}{\rho}.$$

Combining (36) with (14) we are led to

$$(37)\qquad \|\nabla u_{\varepsilon_n}\|_{L^\infty(B(a_j,\rho)\setminus B(a_j,\rho/2))} \le \frac{C}{\rho}.$$

Going back to (35) we obtain

$$(38)\qquad U \le C\rho^2.$$

The desired estimate (10), with $N_2(\rho) = N_4(\rho)$, follows from (29), (32) and (38).

VIII.2. The vanishing gradient property and its various forms

Let $u_\star$ be as in Chapter VI. Recall (see Theorem VII.4) that near each singularity a_j we may write

$$u_\star(z) = \frac{z-a_j}{|z-a_j|}e^{iH_j(z)}$$

where H_j is a harmonic function in some neighborhood of a_j, satisfying

$$(39)\qquad \nabla H_j(a_j) = 0, \quad \forall j.$$

We shall now prove that the vanishing gradient property (39) may also be derived as a consequence of the fact that the configuration $a = (a_j)$ minimizes the renormalized energy W (see Theorem VIII.1). Indeed, (39) is exactly equivalent, as we are going to see, to the assertion that $a = (a_j)$ is a critical point of W, i.e.,

$$(40)\qquad \nabla W(a) = 0.$$

We need some preliminary results. Recall that, given any configuration $b = (b_1, b_2, \dots, b_n)$ with associated integers $(d_1, d_2, \dots, d_n)$ in $\mathbb{Z}$, we have introduced in Section I.3 the solution Φ_0 of the problem

$$\text{(41)} \qquad \begin{cases} \Delta\Phi_0 = \sum_{j=1}^n 2\pi d_j \delta_{b_j} & \text{in } G, \\ \dfrac{\partial\Phi_0}{\partial\nu} = g \times g_\tau & \text{on } \partial G \end{cases}$$

and the corresponding canonical harmonic map u_0 satisfying

$$\text{(42)} \qquad \begin{cases} u_0 \times \dfrac{\partial u_0}{\partial x_1} = -\dfrac{\partial\Phi_0}{\partial x_2} & \text{in } \Omega = G \setminus \bigcup_j \{b_j\}, \\ u_0 \times \dfrac{\partial u_0}{\partial x_2} = \dfrac{\partial\Phi_0}{\partial x_1} & \text{in } \Omega. \end{cases}$$

Moreover (see Corollary I.2) we have, near each b_j,

$$\text{(43)} \qquad u_0(z) = \left(\frac{z - b_j}{|z - b_j|}\right)^{d_j} e^{iH_j(z)}$$

where H_j is a smooth harmonic function in some neighborhood of b_j.

Theorem VIII.2. *Set*

$$\text{(44)} \qquad S_j(x) = \Phi_0(x) - d_j \log|x - b_j|,$$

so that S_j is a smooth harmonic function near b_j. Then S_j and H_j are harmonic conjugates, i.e.,

$$\text{(45)} \qquad \begin{cases} \dfrac{\partial H_j}{\partial x_1} = -\dfrac{\partial S_j}{\partial x_2}, \\ \dfrac{\partial H_j}{\partial x_2} = \dfrac{\partial S_j}{\partial x_1}. \end{cases}$$

Proof. For simplicity we take $b_j = 0$, and rewriting (43) using polar coordinates we have

$$u_0(z) = e^{id_j\theta} e^{iH_j},$$

i.e.,

$$u_0 = (\cos(H_j + d_j\theta), \sin(H_j + d_j\theta)).$$

Hence we find

$$\text{(46)} \qquad \begin{cases} u_0 \times \dfrac{\partial u_0}{\partial x_1} = \dfrac{\partial H_j}{\partial x_1} + d_j \dfrac{\partial\theta}{\partial x_1}, \\ u_0 \times \dfrac{\partial u_0}{\partial x_2} = \dfrac{\partial H_j}{\partial x_2} + d_j \dfrac{\partial\theta}{\partial x_2}. \end{cases}$$

Combining (42), (44) and (46) we obtain

$$(47)\qquad \begin{cases} \dfrac{\partial H_j}{\partial x_1} + d_j \dfrac{\partial \theta}{\partial x_1} = -\dfrac{\partial S_j}{\partial x_2} - d_j \dfrac{\partial}{\partial x_2} \log|x| \\ \dfrac{\partial H_j}{\partial x_2} + d_j \dfrac{\partial \theta}{\partial x_2} = \dfrac{\partial S_j}{\partial x_1} + d_j \dfrac{\partial}{\partial x_1} \log|x|. \end{cases}$$

Finally, we note that the functions θ and $\log|x|$ are harmonic conjugates, i.e.,

$$(48)\qquad \begin{cases} \dfrac{\partial \theta}{\partial x_1} = -\dfrac{\partial}{\partial x_2} \log|x| \\ \dfrac{\partial \theta}{\partial x_2} = \dfrac{\partial}{\partial x_1} \log|x| \end{cases}$$

and thus we deduce (45) from (47) and (48).

Recall that the renormalized energy $W(b)$ was defined in Theorem I.7 where we found

$$(49)\qquad W(b) = -\pi \sum_{i \neq j} d_i d_j \log|b_i - b_j| + \frac{1}{2} \int_{\partial G} \Phi_0 (g \times g_\tau) \\ -\pi \sum_{i=1}^{n} d_i \, R_0(b_i)$$

and

$$(50)\qquad R_0(x) = \Phi_0(x) - \sum_{i=1}^{n} d_i \log|x - b_i|.$$

In the next result we present a simple formula for computing the differential of W considered as a function of $b = (b_1, b_2, \ldots, b_n) \in G^n$.

Theorem VIII.3. *We have*

$$(51)\quad DW(b) = \\ -2\pi \left[d_1 \left(\frac{\partial S_1}{\partial x_1}(b_1), \frac{\partial S_1}{\partial x_2}(b_1) \right), \ldots, d_n \left(\frac{\partial S_n}{\partial x_1}(b_n), \frac{\partial S_n}{\partial x_2}(b_n) \right) \right] = \\ 2\pi \left[d_1 \left(-\frac{\partial H_1}{\partial x_2}(b_1), \frac{\partial H_1}{\partial x_1}(b_1) \right), \ldots, d_n \left(-\frac{\partial H_n}{\partial x_2}(b_n), \frac{\partial H_n}{\partial x_1}(b_n) \right) \right].$$

Before proving Theorem VIII.3 we deduce a simple consequence:

Corollary VIII.1. *The property that $b = (b_1, b_2, \dots, b_n)$ is a critical point of W has several equivalent forms:*

$$\nabla S_j(b_j) = 0 \quad \forall j \tag{52}$$

or

$$\nabla H_j(b_j) = 0 \quad \forall j \tag{53}$$

or

$$\nabla R_0(b_j) + \sum_{i \neq j} d_i \frac{(b_j - b_i)}{|b_j - b_i|^2} = 0 \quad \forall j. \tag{54}$$

Note that the last condition comes from the fact (see (44) and (50)) that, for each j,

$$R_0(x) = S_j(x) - \sum_{i \neq j} d_i \log |x - b_i| \tag{55}$$

and thus

$$\nabla R_0(x) = \nabla S_j(x) - \sum_{i \neq j} d_i \frac{(x - b_i)}{|x - b_i|^2}$$

Proof of Theorem VIII.3. For the convenience of the reader we first describe the proof in the case $n = d = 1$, which is particularly simple. It is useful to introduce more precise notations. Given $y \in G$ let $\Phi(x, y)$ be the solution of

$$\begin{aligned} \Delta\Phi &= 2\pi\, \delta_y && \text{in } G, \\ \frac{\partial\Phi}{\partial\nu} &= g \times g_\tau && \text{on } \partial G \end{aligned}$$

with the normalization condition

$$\int_{\partial G} \Phi(\sigma, y) d\sigma = 0.$$

Note that $\Phi(x, y)$ is well defined for $x \neq y$. By analogy with (50) we introduce

$$R(x, y) = \Phi(x, y) - \log |x - y|, \tag{56}$$

so that $R(x, y)$ is well defined, even for $x = y$, and it is smooth on $G \times G$.

Given two points $b \neq \tilde{b}$ in G we have

$$\begin{aligned} 2\pi\,\Phi(\tilde{b}, b) &= \int_G \Phi(x, b)\Delta\Phi(x, \tilde{b}) \\ &= \int_{\partial G} \Phi(\sigma, b)(g \times g_\tau)d\sigma - \int_G \nabla\Phi(x, b)\nabla\Phi(x, \tilde{b}) \\ &= \int_{\partial G} \left(\Phi(\sigma, b) - \Phi(\sigma, \tilde{b})\right)(g \times g_\tau)d\sigma + 2\pi\Phi(b, \tilde{b}) \end{aligned}$$

and therefore

$$(57) \qquad 2\pi\left(\Phi(b, \tilde{b}) - \Phi(\tilde{b}, b)\right) = \int_{\partial G} \left(\Phi(\sigma, \tilde{b}) - \Phi(\sigma, b)\right)(g \times g_\tau)d\sigma.$$

Using (56) we deduce from (57) that

$$(58) \qquad 2\pi\left(R(b, \tilde{b}) - R(\tilde{b}, b)\right) = \int_{\partial G} \left(\Phi(\sigma, \tilde{b}) - \Phi(\sigma, b)\right)(g \times g_\tau)d\sigma.$$

Differentiating (58) with respect to b (for fixed $\tilde{b}$) yields

$$(59) \qquad 2\pi\left(R_x(b, \tilde{b}) - R_y(\tilde{b}, b)\right) = -\int_{\partial G} \Phi_y(\sigma, b)(g \times g_\tau)d\sigma.$$

Since R is smooth in $G \times G$ we may take $\tilde{b} = b$ and thus we find

$$(60) \qquad 2\pi\left(R_x(b, b) - R_y(b, b)\right) = -\int_{\partial G} \Phi_y(\sigma, b)(g \times g_\tau)d\sigma.$$

Recall that (see (49)) in our special case

$$W(b) = \frac{1}{2}\int_{\partial G} \Phi(\sigma, b)(g \times g_\tau)d\sigma - \pi\, R(b, b)$$

and thus

$$(61) \qquad DW(b) = \frac{1}{2}\int_{\partial G} \Phi_y(\sigma, b)(g \times g_\tau)d\sigma - \pi R_x(b, b) - \pi R_y(b, b).$$

Combining (60) and (61) we are led to

$$DW(b) = -2\pi R_x(b, b).$$

Since in this special case we have (see (44)),

$$S_1(x) = \Phi(x, b) - \log|x - b| = R(x, b)$$

it follows that

$$\nabla S_1(b) = R_x(b,b),$$

and therefore

$$DW(b) = -2\pi \nabla S_1(b),$$

which is the desired conclusion.

We now turn to the general case. Since our purpose is to compute the differential of W as a function of $b_1, b_2, \ldots, b_n$ we shall fix all the points b_i except one of them—say b_j—which will vary and we will denote it b for simplicity. Given $y \in G$ we denote by $\Phi(x,y)$ the solution of

$$\Delta\Phi = 2\pi \sum_{i\neq j} d_i \delta_{b_i} + 2\pi d_j \delta_y \quad \text{in } G,$$

$$\frac{\partial \Phi}{\partial \nu} = g \times g_\tau \qquad \text{on } \partial G$$

with the normalization condition

$$\int_{\partial G} \Phi(\sigma, y) d\sigma = 0.$$

We also introduce

$$\Psi(x,y) = \Phi(x,y) - \sum_{i\neq j} d_i \log|x - b_i|$$

and

$$R(x,y) = \Psi(x,y) - d_j \log|x-y|,$$

so that $R(x,y)$ is well defined, even for $x = y$, and it is smooth on $G \times G$. Note that

$$\begin{cases} \Delta\Psi = 2\pi d_j \delta_y & \text{in } G, \\ \dfrac{\partial \Psi}{\partial \nu} = g \times g_\tau - \sum\limits_{i\neq j} d_i \dfrac{\partial}{\partial \nu} \log|x - b_i| \equiv h & \text{on } \partial G. \end{cases}$$

Given two points $b \neq \tilde{b}$ in G we have, as above,

$$2\pi d_j \left(\Psi(b,\tilde{b}) - \Psi(\tilde{b}, b)\right) = \int_{\partial G} \left(\Psi(\sigma,\tilde{b}) - \Psi(\sigma, b)\right) h(\sigma) d\sigma \tag{62}$$

and consequently

$$2\pi d_j \left(R(b,\tilde{b}) - R(\tilde{b}, b)\right) = \int_{\partial G} \left(\Phi(\sigma,\tilde{b}) - \Phi(\sigma, b)\right) h(\sigma) d\sigma. \tag{63}$$

Next, note that the function

$$\zeta(x) = \Phi(x,\tilde{b}) - \Phi(x,b) = R(x,\tilde{b}) - R(x,b) + d_j \log \frac{|x-\tilde{b}|}{|x-b|}$$

satisfies

$$(64) \qquad \begin{cases} \Delta\zeta = 2\pi d_j(\delta_{\tilde{b}} - \delta_b) & \text{in } G, \\ \dfrac{\partial \zeta}{\partial \nu} = 0 & \text{on } \partial G. \end{cases}$$

Multiplying (64) by $\sum_{i\neq j} d_i \log|x-b_i|$ and integrating yields

$$(65) \qquad 2\pi \sum_{i\neq j} d_i \zeta(b_i) - \int_{\partial G} \zeta \frac{\partial}{\partial \nu}\left(\sum_{i\neq j} d_i \log|x-b_i|\right) = 2\pi \sum_{i\neq j} d_i d_j \log \frac{|\tilde{b}-b_i|}{|b-b_i|}.$$

Inserting (65) in (63) we obtain

$$(66) \quad 2\pi d_j \left(R(b,\tilde{b}) - R(\tilde{b},b)\right) = \int_{\partial G} \left(\Phi(\sigma,\tilde{b}) - \Phi(\sigma,b)\right)(g\times g_\tau) - 2\pi \sum_{i\neq j} d_i \left(R(b_i,\tilde{b}) - R(b_i,b)\right).$$

Differentiating (66) with respect to b (for fixed $\tilde{b}$) we find, taking $\tilde{b}=b$,

$$(67) \quad 2\pi d_j \left(R_x(b,b) - R_y(b,b)\right) = -\int_{\partial G} \Phi_y(\sigma,b)(g\times g_\tau)d\sigma + 2\pi \sum_{i\neq j} d_i R_y(b_i,b).$$

Finally, we recall that the renormalized energy has the form

$$W(b) = -2\pi \sum_{i\neq j} d_i d_j \log|b_i - b| - \pi \sum_{\substack{k\neq \ell \\ k\neq j, \ell\neq j}} d_k d_\ell \log|b_k - b_\ell| + \frac{1}{2}\int_{\partial G} \Phi(\sigma,b)(g\times g_\tau)d\sigma - \pi \sum_{i\neq j} d_i R(b_i,b) - \pi d_j R(b,b).$$

Hence, we obtain

$$W_b(b) = -2\pi \sum_{i\neq j} d_i d_j \frac{b-b_i}{|b-b_i|^2} + \frac{1}{2}\int_{\partial G} \Phi_y(\sigma,b)(g\times g_\tau)d\sigma - \pi \sum_{i\neq j} d_i R_y(b_i,b) - \pi d_j \left(R_x(b,b) + R_y(b,b)\right).$$

Using (67) we see that

$$W_b(b) = -2\pi \sum_{i \neq j} d_i d_j \frac{b - b_i}{|b - b_i|^2} - 2\pi d_j R_x(b, b). \tag{68}$$

On the other hand, we have

$$S_j(x) = \Phi(x, b) - d_j \log |x - b| = R(x, b) + \sum_{i \neq j} d_i \log |x - b_i|$$

and consequently

$$W_b(b) = -2\pi d_j \nabla S_j(b)$$

which yields (51).

Alternative proof of Theorem VIII.3.

This approach relies on the following representation of DW, involving the Hopf differential ω associated with the canonical harmonic map u_0,

$$\omega = \left|\frac{\partial u_0}{\partial x_1}\right|^2 - \left|\frac{\partial u_0}{\partial x_2}\right|^2 - 2i \frac{\partial u_0}{\partial x_1} \cdot \frac{\partial u_0}{\partial x_2},$$

where u_0 is as above (see (42)).

Theorem VIII.4. *Let $B = ((\alpha_1, \beta_1), \dots, (\alpha_n, \beta_n)) \in (\mathbb{R}^2)^n$ be a variation of the configuration b. Let $\sigma > 0$ sufficiently small such that all the discs $B(b_j, \sigma)$ are disjoint, then*

$$DW(b)(B) = \operatorname{Im} \sum_{j=1}^{n} \int_{\partial B(b_j, \sigma)} \frac{1}{2} \omega(z)(\alpha_j + i\beta_j)\, dz. \tag{69}$$

In other words, DW is given by the residues of ω around the singularities b_j.

Before proving Theorem VIII.4, we show how to derive Theorem VIII.3 from this result. Using (43) it is easy to see that, near each b_j,

$$\omega(z) = \left(-i \frac{d_j}{z - b_j} + \frac{\partial H_j}{\partial x_1} - i \frac{\partial H_j}{\partial x_2}\right)^2.$$

Hence

$$\operatorname{Im} \int_{\partial B(b_j, \sigma)} \omega(z)(\alpha_j + i\beta_j)\, dz = 4\pi d_j \left(\frac{\partial H_j}{\partial x_1}(b_j)\beta_j - \frac{\partial H_j}{\partial x_2}(b_j)\alpha_j\right)$$

and the conclusion of Theorem VIII.3 follows from (69).

Proof of Theorem VIII.4. For simplicity we will assume that we have only one singularity b with corresponding degree $+1$. Since we are going to vary the point b, it will be important to emphasize the dependence of the canonical harmonic map on b, by denoting u_b and ω_b instead of u_0 and ω.

Let $\sigma > 0$ be such that $B(b, 2\sigma) \subset G$, and $\chi \in C_c^\infty(G; [0,1])$ which satisfies

$$\begin{cases} \chi(x) = 1 & \text{in } B(b,\sigma) \\ \chi(x) = 0 & \text{in } G \setminus B(b, 2\sigma). \end{cases} \tag{70}$$

Given $B = (\alpha, \beta) \in \mathbb{R}^2$, we let

$$X(x) = (X_1, X_2)(x) = B\,\chi(x),$$

and we choose $t \in \mathbb{R}$ sufficiently small to ensure that

$$U_t : x \mapsto x + tX(x) = x + tB\chi(x) \tag{71}$$

is a diffeomorphism from G to G. We set

$$b_t = b + tB$$

and

$$v_{b,t}(x) = u_{b_t} \circ U_t(x). \tag{72}$$

It follows from Corollary I.2 that

$$u_{b_t}(x) = e^{i\varphi_{b_t}(x)} \frac{z - b_t}{|z - b_t|} \quad \text{in } G, \tag{73}$$

and

$$u_b(x) = e^{i\varphi_b(x)} \frac{(z - b)}{|z - b|} \quad \text{in } G, \tag{74}$$

where φ_b and φ_{b_t} are smooth harmonic functions in G, and $z = x_1 + i\,x_2$. Thus, since by (70) and (71) U_t coincides with a translation on $B(b,\sigma)$, (72) implies that

$$v_{b,t}(x) = e^{i\varphi_{b_t}(x+tB)} \frac{z - b}{|z - b|} \quad \text{in } B(b,\sigma)$$

and hence

$$v_{b,t}(x) = u_b(x) e^{iS_t(x)} \quad \text{in } G, \tag{75}$$

where $S_t \in C^\infty(G;\mathbb{R})$ satisfies

(76)
$$\begin{cases} \operatorname{supp}(\Delta S_t) \subset \overline{B(b,2\sigma)} \setminus B(b,\sigma) \\ S_t = 0 \quad \text{on } \partial G. \end{cases}$$

Moreover it is straightforward, because of (72) and (75), that for any $k \in \mathbb{N}$,

(77)
$$\|\Delta S_t\|_{C^k} = O(t),$$

and thus, for any $k \in \mathbb{N}$,

(78)
$$\|S_t\|_{C^k} = O(t).$$

We now have to compute

(79)
$$W_\rho(b_t) = \frac{1}{2}\int_{G\setminus B(b_t,\rho)} |\nabla u_{b_t}|^2(y)\,dy.$$

Note that since

$$u_{b_t}(y) = v_{b,t} \circ U_t^{-1}(y) = v_{b,t}\,(y - t\,B\chi(y)) + o(t),$$

we have, letting $x = U_t^{-1}(y)$,

(80)
$$\begin{aligned} |\nabla u_{b_t}|^2(y) &= |\nabla v_{b,t}|^2(x) \\ &- 2t\left\{\left|\frac{\partial v_{b,t}}{\partial x_1}\right|^2(x)\frac{\partial X_1}{\partial y_1}(y) + \left|\frac{\partial v_{b,t}}{\partial x_2}\right|^2(x)\frac{\partial X_2}{\partial y_2}(y)\right. \\ &\left. + \frac{\partial v_{b,t}}{\partial x_1}\cdot\frac{\partial v_{b,t}}{\partial x_2}(x)\left(\frac{\partial X_1}{\partial y_2} + \frac{\partial X_2}{\partial y_1}\right)(y)\right\} + o(t). \end{aligned}$$

We set $y = U_t(x)$ in (79), and we get, using the fact that $dy = \left(1 + t\operatorname{div}X + O(t^2)\right)dx$,

(81)
$$\begin{cases} 2W_\rho(b_t) = \displaystyle\int_{G\setminus B(b,\rho)} |\nabla v_{b,t}|^2(x)\,dx \\ -t\displaystyle\int_{G\setminus B(b,\rho)}\left\{\left(|\frac{\partial v_{b,t}}{\partial x_1}|^2 - |\frac{\partial v_{b,t}}{\partial x_2}|^2\right)\left(\frac{\partial X_1}{\partial x_1} - \frac{\partial X_2}{\partial x_2}\right)\right. \\ \qquad\left. + 2\,\dfrac{\partial v_{b,t}}{\partial x_1}\dfrac{\partial v_{b,t}}{\partial x_2}\left(\frac{\partial X_1}{\partial x_2} + \frac{\partial X_2}{\partial x_1}\right)\right\}dx + o(t). \end{cases}$$

We claim that

(82)
$$\int_{G\setminus B(b,\rho)} |\nabla v_{b,t}|^2 dx = \int_{G\setminus B(b,\rho)} |\nabla u_b|^2 + O(t^2) + O(\rho),$$

and

$$\left|\frac{\partial v_{b,t}}{\partial x_1}\right|^2 - \left|\frac{\partial v_{b,t}}{\partial x_2}\right|^2 - 2i\,\frac{\partial v_{b,t}}{\partial x_1}\cdot\frac{\partial v_{b,t}}{\partial x_2} = \omega_b + O(t). \tag{83}$$

We then deduce from (82) and (83) that

$$W_\rho(b_t) = W_\rho(b) - t\int_{G\setminus B(b,\rho)} \operatorname{Re}\left[\omega_b \frac{\partial(X_1+iX_2)}{\partial\bar z}\right] dx + O(t^2) + O(\rho). \tag{84}$$

Thus

$$\begin{aligned} W(b_t) - W(b) &= \lim_{\rho\to 0}(W_\rho(b_t) - W_\rho(b)) \\ &= -t\int_G \operatorname{Re}\left[\omega_b\frac{\partial(X_1+iX_2)}{\partial\bar z}\right] dx + O(t^2), \end{aligned} \tag{85}$$

and

$$DW(b)(B) = -\operatorname{Re}\left[\int_G \omega_b\frac{\partial(X_1+iX_2)}{\partial\bar z}\,dx\right]. \tag{86}$$

We remark that if χ tends to the characteristic function of $B(b,\sigma)$ and if $n=(n_1,n_2)$ is the outward normal to $\partial B(b,\sigma)$ we find

$$DW(b)(B) = -\operatorname{Re}\left[\int_{\partial B(b,\sigma)}\frac{1}{2}(-n_1-in_2)\omega_b(\alpha+i\beta)\,d\mu\right],$$

or

$$DW(b)(B) = \operatorname{Im}\left[\int_{\partial B(b,\sigma)}\frac{1}{2}\omega_b(\alpha+i\beta)\,dz\right]. \tag{87}$$

Thus, to complete the proof, it remains to establish (82) and (83).

Proof of (82). From (75) we deduce

$$v_{b,t}\times\nabla v_{b,t} = (u_b\times\nabla u_b) + \nabla S_t$$

which implies

$$\begin{aligned} &\int_{G\setminus B(b,\rho)} |\nabla v_{b,t}|^2 \\ &\qquad = \int_{G\setminus B(b,\rho)} |\nabla u_b|^2 + 2(u_b\times\nabla u_b)\cdot\nabla S_t + |\nabla S_t|^2 \end{aligned} \tag{88}$$

and

$$(89)\quad \int_{G\setminus B(b,\rho)} 2(u_b \times \nabla u_b)\cdot \nabla S_t = \int_{\partial G} 2S_t u_b \times \frac{\partial u_b}{\partial \nu} - \int_{\partial B(b,\rho)} 2S_t\, u_b \times \frac{\partial u_b}{\partial \nu} - \int_{G\setminus B(b,\rho)} 2S_t\, \operatorname{div}(u_b \times \nabla u_b) = -2\int_{\partial B(b,\rho)} S_t\, u_b \times \frac{\partial u_b}{\partial \nu} = O(\rho).$$

The last equality holds because $\frac{\partial u_b}{\partial \nu}$ is bounded on $\partial B(b,\rho)$ by (74) and because of (76). Moreover (78) implies

$$(90)\quad \int_{G\setminus B(b,\rho)} |\nabla S_t|^2 = O(t^2),$$

and (82) follows from (88), (89) and (90).

Proof of (83). This is an easy consequence of (75) and (77).

VIII.3. Construction of critical points of the renormalized energy

The above characterization of critical points of the renormalized energy leads to the following description of these critical points. Let $b = (b_1, \ldots, b_n)$ be a collection of n distinct points in G, and $d = (d_1, \ldots, d_n)$ in $\mathbf{Z}^n$. Assume that for some boundary condition g on ∂G, (b,d) is a critical point of the renormalized energy. Consider u_b, the associated canonical harmonic map from $G \setminus \{b_1, \ldots, b_n\}$ into S^1, and set

$$\omega_b = \left|\frac{\partial u_b}{\partial x_1}\right|^2 - \left|\frac{\partial u_b}{\partial x_2}\right|^2 - 2i\,\frac{\partial u_b}{\partial x_1}\cdot\frac{\partial u_b}{\partial x_2}.$$

Then ω_b satisfies the following properties:

(P 1) ω_b is holomorphic on $G \setminus \{b_1, \ldots, b_n\}$.

(P 2) There exist n positive integers $e_1, \ldots, e_n$ such that near each point b_j we may write

$$\omega_b(z) = -\frac{e_j^2}{(z-b_j)^2} + A_j(z),$$

where A_j is smooth holomorphic in a neighborhood of b_j (and here $e_j = |d_j|$).

(P 3) The order of each zero of ω_b is even. This means that if $\omega_b(z_0) = 0$, then

$$\omega_b(z) = (z - z_0)^{2k} F(z) \qquad \text{near } z_0.$$

for some integer k and some nonvanishing F.

Note that property (P 3) is just a consequence of the fact that on $G \setminus \{b_1, \ldots, b_n\}$ we may write locally

$$u_b(z) = e^{if_b(z)},$$

where f_b is real harmonic, and then

$$\omega_b = 4\left(\frac{\partial f_b}{\partial z}\right)^2,$$

i.e., ω_b is the (complex) square of some holomorphic function.

Thus we are led to define

$$\mathcal{H}(b,e) = \{\omega : G \to \mathbb{C}; \omega \text{ satisfies properties (P 1), (P 2) and (P 3)}\},$$

where $e = \{e_1, e_2, \ldots, e_n\} \in \mathbb{N}^n$.

Conversely, we have

Theorem VIII.5. *Assume that G is simply connected. Let ω be in $\mathcal{H}(b,e)$; then there exist n integers d_j in $\mathbb{Z}$ such that $|d_j| = e_j$, and there exists a boundary condition $g : \partial G \to S^1$, such that (b,d) is a critical point of the renormalized energy with boundary condition g.*

Proof. Given some ω in $\mathcal{H}(b,e)$, we shall construct some holomorphic function f in $G \setminus \{b_1, \ldots, b_n\}$ that is meromorphic on G, such that $f(z)^2 = \omega(z)$ on G. We first have to check that this construction is possible locally. Here, the only difficulties that occur are near the zeroes of ω and near the singularities of ω. But note that condition (P 3) is precisely the necessary and sufficient condition for constructing a holomorphic square root of ω near the zeroes of u.

Moreover near each singularity b_j, we may write

$$\omega(z) = -\frac{e_j^2}{(z-b_j)^2}\left(1 - \frac{A_j(z)(z-b_j)^2}{e_j^2}\right), \tag{91}$$

and ω is the product of $\left(\dfrac{ie_j}{z-b_j}\right)^2$ with a holomorphic function that does not vanish near b_j. Hence the square root of ω exists at least locally.

Then the construction of a global square root f of ω follows from the fact that G is simply connected (f is unique up to $\{1,-1\}$).

We now set

(92) $$F(z) = \int_{z_0}^{z} f(\zeta)d\zeta;$$

note that this function may be multivalued since near each singularity b_j, f has a pole

$$f(z) = \pm\frac{ie_j}{z-b_j} + R_j.$$

Thus, the real part of F is defined up to $2\pi\mathbb{Z}$. Hence

(93) $$u(z) = e^{i\,\mathrm{Re}(F(z))}$$

is a single-valued map, that takes its values into S^1. Moreover, from the fact that ω is holomorphic, it follows that

(94) $$\left(\frac{\partial}{\partial x_1} - i\frac{\partial}{\partial x_2}\right)(\mathrm{Re}(F)) = \frac{\partial F}{\partial z} = f,$$

and thus

(95) $$\left(\frac{\partial\,\mathrm{Re}(F)}{\partial x_1} - i\frac{\partial\,\mathrm{Re}(F)}{\partial x_2}\right)^2 = \omega.$$

Since F is meromorphic, u is a harmonic map from $G\setminus\{b_1,\ldots,b_n\}$ into S^1. From (95) we deduce that the Hopf differential of u is ω. The degree of u around each singularity b_j is $d_j = \pm e_j$.

In conclusion, if $d = (d_1,\ldots,d_n)$, (b,d) is, by construction, a critical point of the renormalized energy with boundary condition $u_{|\partial G}$.

Example VIII.1. Let

(96) $$\omega = -\frac{1}{(z-1)^2} - \frac{1}{(z+1)^2} = -\frac{2(z-i)(z+i)}{(z^2-1)^2}.$$

Choose any simply connnected domain G that does not contain i and $-i$ (the zeroes of ω). Then ω belongs to $\mathcal{H}((1,-1),(1,1))$. It is easy to verify that the associated harmonic map u has two singularities of degree ±1: one at the point 1, the other at the point -1, with opposite degrees.

VIII.4. The case $G = B_1$ and $g(\theta) = e^{i\theta}$

Throughout this Section we assume that $G = B_1$ is the unit disc and that

(97) $$g(z) = z \quad \text{on } \partial G;$$

in other words $g(\theta) = e^{i\theta}$. Let u_ε be a minimizer for

(98) $$\min_{u\in H^1_g}\left\{\frac{1}{2}\int_G |\nabla u|^2 + \frac{1}{4\varepsilon^2}\int_G (|u|^2-1)^2\right\}$$

where $H^1_g = \{u \in H^1(G;\mathbb{C});\ u = g \text{ on } \partial G\}$.

Our main result is

Theorem VIII.6. *We have*

(99) $$u_\varepsilon(x) \longrightarrow \frac{x}{|x|} = e^{i\theta} \quad \text{in } C^k_{\mathrm{loc}}(G \setminus \{0\})\ \forall k;$$

The convergence also holds in $C^{1,\alpha}_{\mathrm{loc}}(\overline{G} \setminus \{0\})$ $\forall \alpha < 1$.

Proof. From Theorems VI.1 and VI.2 we know (since $\deg(g, \partial G) = +1$) that there is one point $a \in G$ and a subsequence such that

$$u_{\varepsilon_n} \to u_\star \quad \text{in } C^k_{\mathrm{loc}}(G \setminus \{a\}) \text{ and } C^{1,\alpha}_{\mathrm{loc}}(\bar{G} \setminus \{a\}).$$

We first identify a. We claim that

(100) $$a = 0.$$

Proof of (100). Recall that $u_\star$ is the canonical harmonic map associated to the singularity a (see Theorem VII.1). Thus, by Corollary I.2 we know that $u_\star$ has the form

(101) $$u_\star(z) = e^{i\varphi(z)} \frac{(z-a)}{|z-a|}$$

where φ satisfies

(102) $$\begin{cases} \Delta\varphi = 0 & \text{in } G \\ \varphi = \varphi_0 & \text{on } \partial G. \end{cases}$$

and, by (101) and (97),

(103) $$e^{i\varphi_0(z)} = z \frac{|z-a|}{(z-a)} \quad \text{on } \partial G.$$

Suppose by contradiction that $a \neq 0$. Without loss of generality we may assume that

(104) $$0 < a < 1.$$

Using polar coordinates in (103) we write

$$z = e^{i\theta} \qquad \text{with} \quad \theta \in [-\pi, +\pi],$$

and

$$e^{i\varphi_0(\theta)} = \frac{z(\bar{z}-a)}{|z-a|} = \frac{(1 - a\cos\theta) - ai\sin\theta}{(1 - 2a\cos\theta + a^2)^{1/2}}.$$

Thus

$$\cos\varphi_0(\theta) = \frac{1 - a\cos\theta}{(1 - 2a\cos\theta + a^2)^{1/2}} > 0 \quad \forall\theta \in [-\pi, +\pi]$$

and

$$\sin\varphi_0(\theta) = -\frac{a\sin\theta}{(1 - 2a\cos\theta + a^2)^{1/2}}.$$

It follows that

$$(105) \qquad \begin{cases} \varphi_0(\theta) > 0 & \forall\theta \in (-\pi, 0), \\ \varphi_0(\theta) < 0 & \forall\theta \in (0, +\pi). \end{cases}$$

We solve (102) explicitly using Poisson's formula (see e.g., F. Treves [1]) and we find, in polar coordinates,

$$\varphi(r,\theta) = \frac{(1 - r^2)}{2\pi}\int_{-\pi}^{+\pi} \frac{\varphi_0(\sigma)}{1 - 2r\cos(\theta - \sigma) + r^2}d\sigma.$$

In particular, we deduce that

$$\frac{\partial\varphi}{\partial\theta}(r,\theta) = -\frac{r(1 - r^2)}{\pi}\int_{-\pi}^{+\pi} \frac{\varphi_0(\sigma)\sin(\theta - \sigma)d\sigma}{(1 - 2r\cos(\theta - \sigma) + r^2)^2}$$

and thus

$$\frac{\partial\varphi}{\partial\theta}(a,0) = \frac{a(1 - a^2)}{\pi}\int_{-\pi}^{+\pi} \frac{\varphi_0(\sigma)\sin\sigma\, d\sigma}{(1 - 2a\cos\sigma + a^2)^2}.$$

In view of (105) we see that

$$\varphi_0(\sigma)\sin\sigma < 0 \quad \forall\sigma \in (-\pi, +\pi), \sigma \neq 0.$$

and therefore

$$(106) \qquad \frac{\partial\varphi}{\partial\theta}(a,0) < 0.$$

On the other hand we know, by Theorem VII.4 (see also Section VIII.2) that

$$\nabla\varphi(a) = 0$$

This contradicts (106) and hence (100) is proved.

Proof of Theorem VIII.6. Going back to (103) with $a = 0$ we see that $\varphi_0 = 0$ on ∂G and thus $\varphi = 0$ on G. It follows from (101) that

$$u_\star(z) = \frac{z}{|z|} \qquad \text{in } G \setminus \{0\}.$$

So far, we have only proved the convergence of a subsequence u_{ε_n} to $u_\star$. The convergence of the full sequence u_ε follows from the uniqueness of the possible limit via a standard argument.

VIII.5. The case $G = B_1$ and $g(\theta) = e^{di\theta}$ with $d \geq 2$

We start with a simple observation about a general domain G and a general boundary condition $g : \partial G \to S^1$.

Theorem VIII.7. *Assume*

$$\varepsilon \geq \frac{1}{\sqrt{\lambda_1}} \tag{107}$$

where λ_1 is the first eigenvalue of $-\Delta$ on G with zero Dirichlet condition. Then there is a unique minimizer for problem (98).

Proof. We write the Ginzburg-Landau energy E_ε as

$$E_\varepsilon(u) = \frac{1}{2}\int_G \left(|\nabla u|^2 - \frac{1}{\varepsilon^2}|u|^2\right) + \frac{1}{4\varepsilon^2}\int_G |u|^4 + \frac{1}{4\varepsilon^2}|G|.$$

Recall that the function

$$\int_G |\nabla u|^2 - \lambda_1 |u|^2$$

is convex and therefore E_ε is strictly convex for $\varepsilon \geq \frac{1}{\sqrt{\lambda_1}}$. In fact, a monotonicity argument shows that the corresponding Euler equation

$$\begin{cases} -\Delta u = \frac{1}{\varepsilon^2} u(1 - |u|^2) & \text{in } G \\ u = g & \text{on } \partial G \end{cases}$$

has a unique solution when (107) holds.

However, if ε is small, problem (98) need not have a unique minimizer. Here is a simple example:

Theorem VIII.8. *Assume $G = B_1$ and*

$$g(\theta) = e^{di\theta} \qquad \text{with } d \geq 2. \tag{108}$$

For $\varepsilon > 0$ sufficiently small there are infinitely many distinct minimizers for problem (98).

Proof. First we observe that there is an S^1-group action on the minimizers of (98). For every $\alpha \in \mathbb{R}$ and every function $u(z)$ set

$$(R_\alpha u)(z) = e^{-di\alpha} u(e^{i\alpha} z).$$

Note that

$$E_\varepsilon(R_\alpha u) = E_\varepsilon(u) \qquad \forall \alpha,\ \forall u,\ \forall \varepsilon$$

and if $g = u_{|\partial G}$ satisfies (108) then

$$R_\alpha u = e^{di\theta} = g \qquad \text{on } \partial G.$$

In order to show that problem (98) has infinitely many distinct minimizers it suffices to find some minimizer that is not invariant under this group action.

We argue by contradiction and assume that there is a sequence $\varepsilon_n \to 0$ such that every minimizer of E_{ε_n} is invariant under R_α. By Theorem VI.1 we may assume (for a further subsequence) that $u_{\varepsilon_n} \to u_\star$. Hence $u_\star$ is also invariant under R_α. Recall that $u_\star$ has precisely d singularities; since $d \geq 2$, one of the singularities is not at the origin. This contradicts the invariance of $u_\star$ under R_α.

CHAPTER IX

Some additional properties of u_ε

IX.1. The zeroes of u_ε

The main result of this section is:

Theorem IX.1. *Let G be a starshaped domain and let $d = \deg(g, \partial G) > 0$. Then, for every $\varepsilon < \varepsilon_0$ (ε_0 depending only on g and G), u_ε has exactly d zeroes in G and each one is of degree $+1$.*

The main ingredient in the proof of Theorem IX.1 is the following result, which is a particular case of a theorem due to P. Bauman, N. Carlson and D. Phillips [1].

Theorem IX.2. *Let $G = B_1$, and let $g(\theta) = r(\theta)e^{i\varphi_0(\theta)}$, with φ_0 increasing from $[0, 2\pi]$ onto $[0, 2\pi]$ and $r(\theta) > 0$. Then, for every ε there is a unique zero of any minimizer u_ε of E_ε on H^1_g.*

Proof of Theorem IX.1. We argue by contradiction; we assume that there is a sequence u_{ε_n} with $\varepsilon_n \to 0$, such that, for every n, u_{ε_n} does not have d zeroes of degree $+1$. By passing to a subsequence we may assume that

$$u_{\varepsilon_n} \to u_\star \quad \text{in } C^1_{\text{loc}}\left(\overline{G} \setminus \bigcup_j \{a_j\}\right), \tag{1}$$

(see Theorem VI.1). Recall that there are exactly d points in the collection (a_j) (see Theorem VI.2) and that in addition

$$u_\star(z) = \frac{z - a_j}{|z - a_j|} e^{iH_j(z)} \quad \text{near each } a_j, \tag{2}$$

where H_j is a harmonic function in some neighborhood of a_j. We fix $\rho > 0$ sufficiently small so that

$$\|\nabla H_j\|_{L^\infty(B_\rho(a_j))} \leq \frac{1}{2\rho}, \qquad \forall j. \tag{3}$$

Thus, we may write

$$u_\star(z) = e^{i\varphi_j(\theta)} \quad \text{for} \quad z - a_j = \rho e^{i\theta} \tag{4}$$

with

$$\varphi_j(\theta)=\theta+H_j(a_j+\rho e^{i\theta}).$$

Hence

(5) $$\frac{d\varphi_j}{d\theta}\geq\frac{1}{2}.$$

Using (1) we may write for n sufficiently large

(6) $$u_{\varepsilon_n}(z)=r_n(\theta)e^{i\varphi_{j,n}(\theta)}\quad\text{for } z-a_j=\rho e^{i\theta}$$

with $r_n(\theta)>0$.

Moreover

(7) $$\begin{cases} r_n(\theta)\to 1 & \text{in } C([0,2\pi]) \\ \varphi_{j,n}(\theta)\to\varphi_j(\theta) & \text{in } C^1([0,2\pi]). \end{cases}$$

Combining (5) and (7) we may assume that, for n sufficiently large,

(8) $$\frac{d\varphi_{j,n}}{d\theta}\geq\frac{1}{4}.$$

Applying Theorem IX.2 we deduce that u_{ε_n} has exactly one zero of degree $+1$ in each disc $B_\rho(a_j)$. On the other hand we also know from (1) that, for n sufficiently large,

$$|u_{\varepsilon_n}|\geq\frac{1}{2}\quad\text{in } \overline{G}\setminus\bigcup_j B_\rho(a_j).$$

Thus u_{ε_n} has exactly d zeroes of degree $+1$. Contradiction.

IX.2. The limit of $\{E_\varepsilon(u_\varepsilon)-\pi\, d|\log\varepsilon|\}$ as $\varepsilon\to 0$

The main result of this section is:

Theorem IX.3. *Assume G is starshaped, then*

(9) $$\lim_{\varepsilon\to 0}\{E_\varepsilon(u_\varepsilon)-\pi d|\log\varepsilon|\}=\operatorname*{Min}_{G^d} W+d\gamma,$$

where γ is some universal constant.

Before proving Theorem IX.3 we derive an easy consequence of Theorem V.3:

Lemma IX.1. *Assume $G = B_1$ and $g(\theta) = e^{i\theta}$. Then,*

$$(10) \qquad \gamma \equiv \lim_{\varepsilon\to 0} \{E_\varepsilon(u_\varepsilon) - \pi|\log \varepsilon|\} \quad \textit{exists and is finite.}$$

Proof of Lemma IX.1. Recall that

$$E_\varepsilon(u_\varepsilon) = I(\varepsilon)$$

and that $t \mapsto I(t) + \pi \log t$ is nondecreasing (see Lemma III.1). Hence

$$\lim_{\varepsilon\to 0} \{I(\varepsilon) - \pi|\log \varepsilon|\} \quad \text{exists in } [-\infty, +\infty).$$

By Theorem V.3 we know that this limit is finite.

Proof of Theorem IX.3. We shall use the estimates of Lemma VIII.1 and Lemma VIII.2 in conjunction with Lemma IX.1. First, by Lemma VIII.1, we know that there is some ρ_0 such that for every $\rho < \rho_0$ and every $\varepsilon > 0$, we have

$$(11) \qquad E_\varepsilon(u_\varepsilon) + \pi d \log \varepsilon \le d\left[I\left(\frac{\varepsilon}{\rho}\right) + \pi \log\left(\frac{\varepsilon}{\rho}\right)\right] + \operatorname*{Min}_{G^d} W + O(\rho).$$

Hence

$$\limsup_{\varepsilon\to 0} \{E_\varepsilon(u_\varepsilon) - \pi d|\log \varepsilon|\} \le d\gamma + \operatorname*{Min}_{G^d} W + O(\rho).$$

Since this is true for any $\rho < \rho_0$ we have

$$(12) \qquad \limsup_{\varepsilon\to 0} \{E_\varepsilon(u_\varepsilon) - \pi d|\log \varepsilon|\} \le d\gamma + \operatorname*{Min}_{G^d} W.$$

We now claim that

$$(13) \qquad \liminf_{\varepsilon\to 0} \{E_\varepsilon(u_\varepsilon) - \pi d|\log \varepsilon|\} \ge d\gamma + \operatorname*{Min}_{G^d} W.$$

We argue by contradiction; suppose that

$$\liminf_{\varepsilon\to 0} \{E_\varepsilon(u_\varepsilon) - \pi d|\log \varepsilon|\} < d\gamma + \operatorname*{Min}_{G^d} W.$$

Hence, there is a sequence $\varepsilon_n \to 0$ such that

$$(14) \qquad \lim_{n\to +\infty} \{E_{\varepsilon_n}(u_{\varepsilon_n}) - \pi d|\log \varepsilon_n|\} = \ell < d\gamma + \operatorname*{Min}_{G^d} W.$$

By passing to a subsequence we may assume that

$$u_{\varepsilon_n} \longrightarrow u_\star,$$

as in Theorem VI.1. We are now in a position to apply Lemma VIII.2. It says that given any $\rho(<\rho_1)$, there exists an integer $N(\rho)$ such that for any $n \geq N(\rho)$,

$$E_{\varepsilon_n}(u_{\varepsilon_n}) + \pi d \log \varepsilon_n \geq d\left[I\left(\frac{\varepsilon_n}{\rho}\right) + \pi \log\left(\frac{\varepsilon_n}{\rho}\right)\right] + \operatorname*{Min}_{G^d} W + O(\rho^2). \tag{15}$$

Hence

$$\ell \geq d\gamma + \operatorname*{Min}_{G^d} W + O(\rho^2).$$

Since this is true for any ρ, we have a contradiction with (14).

IX.3. $\int_G |\nabla|u_\varepsilon||^2$ remains bounded as $\varepsilon \to 0$

The main result of this section is:

Theorem IX.4. *Assume G is starshaped, then*

$$\int_G |\nabla|u_\varepsilon||^2 \leq C \quad \text{as } \varepsilon \to 0. \tag{16}$$

Proof. We return to the proof of Theorem V.2. Instead of (V.8) we have, in fact, a better estimate:

$$\int_{\Omega_j} |\nabla u_{\varepsilon_n}|^2 \geq \int_{\Omega_j} |\nabla|u_{\varepsilon_n}||^2 + \int_{\Omega_j} |\nabla v_{\varepsilon_n}|^2 - C. \tag{17}$$

As in the proof of Theorem V.1 we deduce with the help of (V.7) that

$$\int_{\Omega_j} |\nabla|u_{\varepsilon_n}||^2 \leq C. \tag{18}$$

On the other hand we have

$$\int_{B(x_i,\lambda\varepsilon)} |\nabla u_\varepsilon|^2 \leq C \tag{19}$$

since $\|\nabla u_\varepsilon\|_{L^\infty} \leq C/\varepsilon$.

Combining (18) and (19) we are led to

$$\int_{B(a_j,\eta)} |\nabla|u_{\varepsilon_n}||^2 \leq C. \tag{20}$$

The desired conclusion follows from (20) and Theorem V.1.

IX.4. The bad discs revisited

In Chapter IV we have constructed a family of (modified) bad discs $B(x_i, \lambda\varepsilon)_{i\in J}$ with $x_i = x_i^{\varepsilon_n}$ and J independent of ε, card $J = N_1$, such that

$$(21) \qquad |u_{\varepsilon_n}(x)| \geq \frac{1}{2} \qquad \forall x \in G \setminus \bigcup_{i\in J} B(x_i, \lambda\varepsilon),$$

$$(22) \qquad |x_i - x_j| \geq 8\lambda\varepsilon \qquad \forall i, j \in J, i \neq j,$$

$$(23) \qquad \frac{1}{\varepsilon_n^2} \int_{B(x_i, 2\lambda\varepsilon)} \left(|u_{\varepsilon_n}|^2 - 1\right)^2 \geq \mu_0,$$

$$(24) \qquad d_i = \deg\left(u_{\varepsilon_n}, \partial B(x_i, \lambda\varepsilon_n)\right).$$

Recall that

$$x_i^{\varepsilon_n} \to \ell_i \qquad \forall i \in J$$

and that the collection of distinct points in $(\ell_i)_{i\in J}$ coincides with $\{a_1, a_2, \ldots, a_d\}$. We have introduced, for every $j = 1, 2, \ldots, d$,

$$\Lambda_j = \left\{i \in J; x_i^\varepsilon \to a_j\right\}$$

and we have proved that, for every j,

$$(25) \qquad \kappa_j = \deg\left(u_{\varepsilon_n}, \partial B(a_j, \eta/2)\right) = \deg\left(u_*, \partial B(a_j, \eta/2)\right) = \sum_{i\in\Lambda_j} d_i = +1.$$

Theorem IX.5. *We have*

$$d_i \in \{0, +1\} \qquad \forall i \in J.$$

Moreover, for every $j = 1, 2, \ldots, d$, there is exactly one $i \in \Lambda_j$ such that $d_i = +1$.

Proof. From (25) we see that, for every j, there is at least one index $i \in \Lambda_j$ such that $d_i \neq 0$. In particular, u_{ε_n} has at least one zero near each a_j.

We have to show that, for every j, there is precisely one $i \in \Lambda_j$ such that $d_i \neq 0$. Suppose, by contradiction, that for some j, there are two indices $i_1, i_2 \in \Lambda_j$ such that $d_{i_1} \neq 0$ and $d_{i_2} \neq 0$. Then u_{ε_n} would have a zero both in $B(x_{i_1}, \lambda\varepsilon_n)$ and in $B(x_{i_2}, \lambda\varepsilon_n)$. Hence, u_{ε_n} would have at least $(d+1)$ zeroes in G and this is impossible by Theorem IX.1.

Theorem IX.6. *For every $j = 1, 2, \dots, d$, we have*

$$\limsup_{n\to\infty} \frac{|x_k^{\varepsilon_n} - x_\ell^{\varepsilon_n}|}{\varepsilon_n} < \infty \qquad \forall k, \ell \in \Lambda_j. \tag{26}$$

Proof. We argue by contradiction. Assume that, for some j, there exist $k \in \Lambda_j$ and $\ell \in \Lambda_j$ such that (passing to a further subsequence)

$$\lim_{n\to\infty} \frac{|x_k^{\varepsilon_n} - x_\ell^{\varepsilon_n}|}{\varepsilon_n} = +\infty. \tag{27}$$

Set $x_{k,n} = x_k^{\varepsilon_n}$ and

$$\rho_n = \frac{|x_{k,n} - x_{\ell,n}|}{\varepsilon_n},$$
$$v_{k,n}(y) = u_{\varepsilon_n}(x_{k,n} + \varepsilon_n y),$$
$$v_{\ell,n}(y) = u_{\varepsilon_n}(x_{\ell,n} + \varepsilon_n y).$$

These functions are defined for $y \in B(0, \rho_n/3)$ and they satisfy

$$-\Delta v = v(1 - |v|^2) \quad \text{in } B(0, \rho_n/3). \tag{28}$$

Since $B(x_{k,n}, \varepsilon_n\rho_n/3) \cap B(x_{\ell,n}, \varepsilon_n\rho_n/3) = \emptyset$ we have, by Theorem VII.2,

$$\begin{aligned}\frac{1}{4\varepsilon_n^2}\int_{B(x_{k,n},\varepsilon_n\rho_n/3)} (|u_{\varepsilon_n}|^2 - 1)^2 + \frac{1}{4\varepsilon_n^2}\int_{B(x_{\ell,n},\varepsilon_n\rho_n/3)} \left(|u_{\varepsilon_n}|^2 - 1\right)^2 \\ \le \frac{\pi}{2} + o(1).\end{aligned} \tag{29}$$

Changing variables we obtain

$$\int_{B(0,\rho_n/3)} \left(|v_{k,n}|^2 - 1\right)^2 + \int_{B(0,\rho_n/3)} \left(|v_{\ell,n}|^2 - 1\right)^2 \le 2\pi + o(1). \tag{30}$$

As $n \to \infty$ we have

$$v_{k,n} (\text{resp. } v_{\ell,n}) \to v_k \text{ (resp. } v_\ell) \text{ in } C^k_{\text{loc}}(\mathbb{R}^2).$$

Both v_k and v_ℓ satisfy

$$-\Delta v = v\left(1 - |v|^2\right) \quad \text{in } \mathbb{R}^2. \tag{31}$$

Applying a result of H. Brezis, F. Merle and T. Rivière [1] (see also Appendix III at the end of the book) we know that every solution of (31) satisfies

$$\int_{\mathbb{R}^2} \left(|v|^2 - 1\right)^2 = 2\pi q^2 \quad \text{with } q = 0, 1, 2, \dots, \infty. \tag{32}$$

Passing to the limit in (30) we find

$$\int_{\mathbf{R}^2} \left(|v_k|^2 - 1\right)^2 + \int_{\mathbf{R}^2} \left(|v_\ell|^2 - 1\right)^2 \leq 2\pi. \tag{33}$$

Combining (32) and (33) we see that one of the integrals (at least) vanishes—say, for example, the integral corresponding to v_k.

On the other hand, going back to (23) we have

$$\int_{B(0,2\lambda)} \left(|v_{k,n}|^2 - 1\right)^2 \geq \mu_0. \tag{34}$$

Passing to the limit in (34) we are led to

$$0 = \int_{B(0,2\lambda)} \left(|v_k|^2 - 1\right)^2 \geq \mu_0.$$

Impossible.

Conclusion: In view of Theorems IX.5 and IX.6 we may still modify further the bad discs. The new family consists of **exactly** d discs $B(x_i^{\varepsilon_n}, \alpha\varepsilon_n)$, $i = 1, 2, \ldots, d$, for some fixed constant α (depending only on g and G), such that

$$|u_{\varepsilon_n}(x)| \geq \frac{1}{2} \quad \forall x \in G \setminus \bigcup_{i=1}^{d} B(x_i^{\varepsilon_n}, \alpha\varepsilon_n),$$

$$\frac{1}{\varepsilon_n^2} \int_{B(x_i^{\varepsilon_n}, \alpha\varepsilon_n)} \left(|u_{\varepsilon_n}|^2 - 1\right)^2 \geq \mu_0 \quad \forall i, \forall n,$$

$$\deg\left(u_{\varepsilon_n}, \partial B(x_i^{\varepsilon_n}, \alpha\varepsilon_n)\right) = +1 \quad \forall i, \forall n$$

and

$$x_i^{\varepsilon_n} \longrightarrow a_i \quad \forall i = 1, 2, \ldots, d.$$

CHAPTER X

Non-minimizing solutions of the Ginzburg-Landau equation

Throughout this chapter, we analyze the behavior as $\varepsilon \to 0$ of solutions v_ε of the Ginzburg-Landau equation:

$$-\Delta v_\varepsilon = \frac{1}{\varepsilon^2} v_\varepsilon (1 - |v_\varepsilon|^2) \quad \text{in } G, \tag{1}$$

$$v_\varepsilon = g \quad \text{on } \partial G. \tag{2}$$

Here v_ε need not be a minimizer of E_ε. We have not investigated the existence of non-minimizing solutions. However, in some cases it is clear that there exist solutions of (1) and (2) that are definitely not minimizers. For example, if $G = B_1$, $g(\theta) = e^{di\theta}$, with $d \geq 2$ there exists a solution of (1)–(2) of the form

$$v_\varepsilon(r, \theta) = e^{di\theta} f_{\varepsilon,d}(r) \tag{3}$$

(see e.g., Appendix II). From the analysis of Section VIII.5 we know that if ε is sufficiently small, v_ε given by (3) is not a minimizer.

Also, here, we do not make any assumption about the degree of g. In particular, the case where $\deg(g, \partial G) = 0$ is of interest (see Open Problem 5 in Chapter XI).

Throughout this chapter we assume that G is starshaped. Our aim here is to show that some of the results presented above for minimizers are still valid. In particular, we will prove that v_{ε_n} converges to some limit $v_\star$ in $C^k_{\text{loc}}(G \setminus \bigcup\{a_j\})$, for a finite set of points a_j. In contrast with the previous situation, the singularities a_j of $v_\star$ need not have degree one (for example, v_ε given by (3) converges to $v_\star = e^{di\theta}$ which has exactly one singularity of degree d). Set $d_j = \deg(v_\star, a_j)$. We will prove that $v_\star$ is the canonical harmonic map associated to (a_j, d_j), in the sense of Section I.3. The location of the singularities a_j is still governed by the renormalized energy W. More precisely, the configuration $a = (a_j)$ is a critical point of W, but it need not be a minimizer of W.

X.1. Preliminary estimates; bad discs and good discs

In this section we present some of the arguments that have been developed for minimizers, but can still be carried over for general solutions of the Ginzburg-Landau equation (1)–(2).

Lemma X.1. *Assume G is starshaped. Then there is a constant C depending only on g and G such that any solution v_ε of (1)–(2) satisfies*

$$\int_{\partial G} \left|\frac{\partial v_\varepsilon}{\partial \nu}\right|^2 + \frac{1}{\varepsilon^2}\int_G (1-|v_\varepsilon|^2) \leq C. \tag{4}$$

The proof is exactly the same as the proof of Theorem III.2. Estimate (4) plays a crucial role in our analysis. **Therefore we assume throughout the rest of this chapter that G is starshaped.**

We also recall

Lemma X.2. *Any solution v_ε of (1)–(2) satisfies*

$$|v_\varepsilon| \leq 1 \qquad \text{in } G, \tag{5}$$

$$|\nabla v_\varepsilon| \leq \frac{C}{\varepsilon} \qquad \text{in } G, \tag{6}$$

where C depends only on g and G.

Estimate (5) follows easily from the maximum principle, and (6) relies for example on Lemma A.2 in the Appendix of F. Bethuel, H. Brezis and F. Hélein [2]. Combining Lemma X.1 and Lemma X.2, we see that Theorem III.3 holds with u_ε replaced by v_ε. Therefore we may carry out the covering argument of Section IV.1 to assert

Lemma X.3. *There exists an integer N depending only on G and g, and a collection of points $(x_i) = (x_i^\varepsilon)$ with $i \in J = J_\varepsilon$ such that*

$$0 \leq \operatorname{card} J_\varepsilon \leq N \tag{7}$$

$$|v_\varepsilon(x)| \geq \frac{1}{2} \qquad \forall x \in G \setminus \cup_{i\in J} B(x_i, \lambda_0\varepsilon) \tag{8}$$

$$\frac{1}{\varepsilon^2}\int_{B(x_i,2\lambda_0\varepsilon)\cap G} (1-|v_\varepsilon|^2)^2 \geq \mu_0, \tag{9}$$

where N, λ_0 and μ_0 are positive constants (depending only on g and G).

The bad discs $B(x_i, \lambda_0\varepsilon)$, with $i \in J$ may intersect. To avoid this unpleasant situation, we replace (as in Section IV.2) the bad discs $B(x_i, \lambda_0\varepsilon)$ by slightly larger discs $B(x_i, \lambda\varepsilon)$, with $i \in J' \subset J$ and $\lambda > \lambda_0$ is a constant depending on g and G, such that

$$B(x_i, 2\lambda\varepsilon) \cap B(x_j, 2\lambda\varepsilon) = \emptyset \quad \text{if } i \neq j.$$

We set

$$\omega_i = B(x_i, \lambda\varepsilon)$$
$$\Omega_\varepsilon = G \setminus \cup_{i\in J'} \omega_i$$
$$\widetilde{G}_\varepsilon = G \setminus \cup_{i\in K} \omega_i,$$

where

$$K = \{i \in J';\ \partial G \cap \omega_i \neq \emptyset\}$$

and

$$L = J' \setminus K = \{i \in J';\ \omega_i \subset G\}.$$

Clearly, we have, by (6),

(10) $$\sum_{i\in L} \int_{\partial\omega_i} |\nabla v_\varepsilon| \leq C$$

and by (6) and (4)

(11) $$\int_{\partial\widetilde{G}_\varepsilon} |\nabla v_\varepsilon| \leq C,$$

where C depends only on g and G.

X.2. Splitting $|\nabla v_\varepsilon|$

In Chapter I, we have related the study of S^1-valued harmonic maps to linear equations. We shall use here a somewhat similar device. Note that, if we write locally, on the set where $|v_\varepsilon| > 0$,

(12) $$v_\varepsilon = \rho_\varepsilon e^{i\psi_\varepsilon}, \qquad \text{with } \rho_\varepsilon = |v_\varepsilon|$$

then (1) transforms into the system:

(13) $$\operatorname{div}(\rho_\varepsilon^2 \nabla\psi_\varepsilon) = 0,$$

(14) $$-\Delta\rho_\varepsilon + \rho_\varepsilon |\nabla\psi_\varepsilon|^2 = \frac{1}{\varepsilon^2}\rho_\varepsilon(1 - \rho_\varepsilon^2).$$

Note, however, that we cannot write (12) globally since ρ_ε vanishes at some points; the corresponding ψ_ε need not be well defined as a single-valued function. To overcome this difficulty we proceed as follows.

Let Φ_ε be the solution of the linear problem

$$\operatorname{div}\left(\frac{1}{\rho_\varepsilon^2}\nabla\Phi_\varepsilon\right) = 0 \qquad \text{in } \Omega_\varepsilon, \tag{15}$$

$$\Phi_\varepsilon = \text{Const.} = C_i \quad \text{on } \partial\omega_i, \text{ for } i \in L, \tag{16}$$

$$\Phi_\varepsilon = 0 \qquad \text{on } \partial\tilde{G}_\varepsilon, \tag{17}$$

$$\int_{\partial\omega_i} \frac{1}{\rho_\varepsilon^2}\frac{\partial\Phi_\varepsilon}{\partial\nu} = 2\pi\delta_i \qquad \text{for } i \in L, \tag{18}$$

where

$$\delta_i = \deg(v_\varepsilon, \partial\omega_i) \quad \text{for } i \in L. \tag{19}$$

We recall that

$$\rho_\varepsilon \geq \frac{1}{2} \quad \text{in } \Omega_\varepsilon \tag{20}$$

by (8), and hence (15) is elliptic and (19) is well defined. Therefore Φ_ε exists and is unique. Moreover, Φ_ε is obtained by minimizing

$$F(\varphi) = \frac{1}{2}\int_{\Omega_\varepsilon} \frac{1}{\rho_\varepsilon^2}|\nabla\varphi|^2 + 2\pi\sum_{i\in L}\delta_i\varphi_{|\partial\omega_i}$$

in the class

$$V = \left\{\varphi \in H^1(\Omega_\varepsilon;\mathbb{R}); \varphi = 0 \text{ on } \partial\tilde{G}_\varepsilon, \text{ and } \varphi = \text{Const. on each } \partial\omega_i\right\}.$$

Whenever there is no confusion, we shall drop the subscript ε. On $\Omega, v/|v|$ is a smooth S^1-valued map, and thus

$$\left(\frac{v}{|v|}\right)_{x_1} \times \left(\frac{v}{|v|}\right)_{x_2} = 0. \tag{21}$$

Rewriting (21), we have

$$\frac{\partial}{\partial x_1}\left[\frac{v}{|v|} \times \left(\frac{v}{|v|}\right)_{x_2}\right] - \frac{\partial}{\partial x_2}\left[\frac{v}{|v|} \times \left(\frac{v}{|v|}\right)_{x_1}\right] = 0.$$

This yields

$$\frac{\partial}{\partial x_1}\left(\frac{1}{\rho^2}v \times v_{x_2}\right) - \frac{\partial}{\partial x_2}\left(\frac{1}{\rho^2}v \times v_{x_1}\right) = 0. \tag{22}$$

By analogy with the proof of Theorem I.1, we set

$$D = \left(\frac{1}{\rho^2} (-v \times v_{x_2} + \Phi_{x_1}), \ \frac{1}{\rho^2} (v \times v_{x_1} + \Phi_{x_2}) \right).$$

Note that

$$\operatorname{div} D = 0 \qquad \text{by (15) and (22)} \tag{23}$$

and

$$\int_{\partial\omega_i} D \cdot \nu = 0 \qquad \text{for } i \in L \tag{24}$$

since

$$2\pi\delta_i = \int_{\partial\omega_i} \frac{v}{|v|} \times \left(\frac{v}{|v|} \right)_\tau = \int_{\partial\omega_i} \frac{1}{\rho^2} v \times v_\tau = \int_{\partial\omega_i} \frac{1}{\rho^2} \frac{\partial\Phi}{\partial\nu}.$$

By Lemma I.1 there is some function H defined in Ω such that

$$D = \left(-\frac{\partial H}{\partial x_2}, \frac{\partial H}{\partial x_1} \right),$$

that is,

$$v \times v_{x_1} + \Phi_{x_2} = \rho^2 H_{x_1}, \tag{25}$$

$$v \times v_{x_2} - \Phi_{x_1} = \rho^2 H_{x_2}. \tag{26}$$

We claim that

$$\operatorname{div}(\rho^2 \nabla H) = 0 \qquad \text{in } \Omega. \tag{27}$$

Indeed

$$\operatorname{div}(\rho^2 \nabla H) = \frac{\partial}{\partial x_1}(v \times v_{x_1}) + \frac{\partial}{\partial x_2}(v \times v_{x_2}),$$

and on the other hand,

$$\frac{\partial}{\partial x_1}(v \times v_{x_1}) + \frac{\partial}{\partial x_2}(v \times v_{x_2}) = v \times \Delta v = 0 \quad \text{in } G \tag{28}$$

by (1).

By (25) and (26), we have

$$|v \times \nabla v| \leq |\nabla\Phi| + |\nabla H| \qquad \text{in } \Omega. \tag{29}$$

Finally, we claim that

$$|\nabla v| \leq |\nabla \rho| + \frac{1}{\rho}|v \times \nabla v|. \tag{30}$$

Indeed if we write locally $v = \rho e^{i\psi}$ we easily see that

$$v \times \nabla v = \rho^2 \nabla \psi \tag{31}$$

and that

$$|\nabla v| \leq |\nabla \rho| + \rho |\nabla \psi|. \tag{32}$$

Combining (31) and (32) we are led to (30). Putting together (29) and (30) we obtain

$$|\nabla v_\varepsilon| \leq 2\Big[|\nabla \Phi_\varepsilon| + |\nabla H_\varepsilon| + |\nabla \rho_\varepsilon|\Big] \qquad \text{in } \Omega_\varepsilon. \tag{33}$$

This estimate will play a crucial role. In what follows, we shall estimate successively $|\nabla \Phi_\varepsilon|$, $|\nabla H_\varepsilon|$ and $|\nabla \rho_\varepsilon|$ in various norms.

X.3. Study of the associated linear problems

We start with some general facts about linear elliptic problems in divergence form. Let G be a smooth, bounded and simply connected domain in $\mathbb{R}^2$, and let ω_i, for $i = 1, \dots, n$ be open, smooth and simply connected subsets of G, with $\omega_i \subset G$ and $\omega_i \cap \omega_j = \emptyset$ for $i \neq j$. Let $\Omega = G \setminus \bigcup_{i=1}^{n} \overline{\omega}_i$. For $i = 1, \dots, n$, let d_i be n numbers in $\mathbb{R}$, and set

$$d = \sum_{i=1}^{n} |d_i|,$$

$$d_0 = \sum_{i=1}^{n} d_i.$$

Let w be a function satisfying

$$\operatorname{div}(a \nabla w) = 0 \qquad \text{in } \Omega \tag{34}$$

where $a : \Omega \to \mathbb{R}$ is a positive function such that

$$\alpha \leq a \leq \alpha^{-1} \tag{35}$$

for some constant $0 < \alpha < 1$.

Lemma X.4. *Assume that w satisfies (34) and*

$$\int_{\partial\omega_i} a\frac{\partial w}{\partial\nu} = 0 \qquad \text{for } i = 1,\dots,n. \tag{36}$$

Then

$$\operatorname*{Sup}_{\Omega} w - \operatorname*{Inf}_{\Omega} w \le \sum_{j=1}^{n}\left[\operatorname*{Sup}_{\partial\omega_j} w - \operatorname*{Inf}_{\partial\omega_j} w\right] + \operatorname*{Sup}_{\partial G} w - \operatorname*{Inf}_{\partial G} w. \tag{37}$$

The proof is exactly the same as the proof of Lemma I.4 (see also Lemma I.3), and we shall omit it.

In what follows we shall present some properties of the solution Φ of the equation

$$\operatorname{div}(a\nabla\Phi) = 0 \qquad \text{in } \Omega, \tag{38}$$

$$\Phi = \text{Const.} = C_i \qquad \text{on } \partial\omega_i, i = 1,2,\dots,n \tag{39}$$

$$\Phi = 0 \qquad \text{on } \partial G \tag{40}$$

$$\int_{\partial\omega_i} a\frac{\partial\Phi}{\partial\nu} = 2\pi d_i \qquad \text{for } i = 1,2,\dots,n, \tag{41}$$

where a satisfies (35).

Lemma X.5. *We have*

$$\operatorname*{Sup}_{\Omega} |\Phi| \le A\frac{d}{\alpha}\left(\operatorname*{Max}_{1\le i\le n} \log\frac{|G|}{|\omega_i|} + 1\right) \tag{42}$$

where A is some universal constant.

Proof. Recall the Trudinger's inequality (see e.g., D. Gilbarg and N. Trudinger [1], Theorem 7.15): there exist two universal constants σ_1 and σ_2 such that

$$\int_G \exp\left(\frac{|u|}{\sigma_1\|\nabla u\|_2}\right)^2 \le \sigma_2|G| \qquad \forall u \in H_0^1(G). \tag{43}$$

We apply (43) with

$$u = \begin{cases} \Phi & \text{in } \Omega, \\ C_i & \text{in } \omega_i. \end{cases}$$

From (43) it follows that

$$\sum_{i=1}^{n} \exp\left(\frac{C_i}{\sigma_1\|\nabla u\|_2}\right)^2 |\omega_i| \le \sigma_2|G|$$

and in particular, for each i,

(44) $$|C_i|^2 \le \sigma_1^2 \|\nabla u\|_2^2 \ \log\left(\frac{\sigma_2|G|}{|\omega_i|}\right).$$

On the other hand, if we multiply (38) by Φ and integrate over Ω we find

$$\alpha \int_\Omega |\nabla\Phi|^2 \le 2\pi \sum_i |C_i\, d_i|$$

and thus

(45) $$\|\nabla u\|_2^2 = \int_\Omega |\nabla\Phi|^2 \le \frac{2\pi}{\alpha}\sum_i |C_i\, d_i| \le \frac{2\pi}{\alpha} d \operatorname*{Max}_i C_i\,.$$

By the maximum principle,

$$\operatorname*{Sup}_\Omega |\Phi| = \operatorname*{Max}_i |C_i|$$

and the conclusion follows easily from (44) and (45).

Lemma X.6. *Assume Φ satisfies (38)–(41). Then we have*

(46) $$\int_{\partial\omega_i} a\left|\frac{\partial\Phi}{\partial\nu}\right| \le 4\pi\, d \quad \textit{for } i = 0, 1, \ldots, n$$

where $\partial\omega_0 = \partial G$.

Proof. By linearity we may always assume that each d_i is zero except one of them, say $d_1 = 1$ and $d_j = 0$ for $j = 2, 3, \ldots, n$.

We claim that

(47) $$\int_{\partial\omega_i} a\left|\frac{\partial\Phi}{\partial\nu}\right| \le 4\pi \qquad \text{for } i = 0, 1, 2, \ldots, n.$$

Proof of (47) for $i = 1$. It is easy to see, as in the proof of Lemma II.2, that

(48) $$\frac{\partial\Phi}{\partial\nu} \ge 0 \qquad \text{on } \partial\omega_1$$

and thus

$$\int_{\partial\omega_1} a\left|\frac{\partial\Phi}{\partial\nu}\right| = \int_{\partial\omega_1} a\,\frac{\partial\Phi}{\partial\nu} = 2\pi.$$

Proof of (47) for $i = 0$. We also know (as in Lemma II.2) that

(49) $$\Phi \le 0 \qquad \text{in } \Omega$$

and therefore (since $\Phi = 0$ on ∂G),

$$\frac{\partial \Phi}{\partial \nu} \geq 0 \qquad \text{on } \partial G. \tag{50}$$

Integrating (38) over Ω we have

$$\int_{\partial G} a \frac{\partial \Phi}{\partial \nu} = \int_{\partial \omega_1} a \frac{\partial \Phi}{\partial \nu} = 2\pi. \tag{51}$$

Proof of (47) for $i \geq 2$. Choose for example $i = 2$ and let ζ be the solution of the linear problem

$$\operatorname{div}(a \nabla \zeta) = 0 \qquad \text{in } \Omega \tag{52}$$

$$\zeta = \text{Const.} = C_i \qquad \text{on } \partial\omega_i \text{ for } i \neq 2 \tag{53}$$

$$\zeta = 0 \qquad \text{on } \partial G \tag{54}$$

$$\zeta = h \qquad \text{on } \partial\omega_2 \tag{55}$$

$$\int_{\partial \omega_i} a \frac{\partial \zeta}{\partial \nu} = 0 \qquad \text{for } i \neq 2 \text{ and } i \neq 0 \tag{56}$$

where $h : \partial\omega_2 \to \mathbb{R}$ is a given function such that

$$|h| \leq 1 \qquad \text{on } \partial\omega_2. \tag{57}$$

Clearly, problem (52)–(56) has a unique solution. Integrating (52) over Ω we have

$$\int_{\partial G} a \frac{\partial \zeta}{\partial \nu} = \int_{\partial \omega_2} a \frac{\partial \zeta}{\partial \nu}. \tag{58}$$

We denote by γ the solution of (52)–(56) corresponding to $h \equiv +1$. By the maximum principle (as in the proof of Lemma II.2) we easily see that

$$0 \leq \gamma \leq 1 \quad \text{in } \Omega \tag{59}$$

and

$$|\zeta| \leq \gamma \quad \text{in } \Omega. \tag{60}$$

By (58) we also have

$$\int_{\partial G} a \frac{\partial \gamma}{\partial \nu} = \int_{\partial \omega_2} a \frac{\partial \gamma}{\partial \nu}.$$

From (60) (and since $\zeta = \gamma = 0$ on ∂G) we obtain

$$\left|\frac{\partial \zeta}{\partial \nu}\right| \leq -\frac{\partial \gamma}{\partial \nu} \quad \text{on } \partial G. \tag{61}$$

Thus

$$(62)\quad \left|\int_{\partial\omega_2} a\frac{\partial\zeta}{\partial\nu}\right| = \left|\int_{\partial G} a\frac{\partial\zeta}{\partial\nu}\right| \le \int_{\partial G} a\left|\frac{\partial\zeta}{\partial\nu}\right| \le -\int_{\partial G} a\frac{\partial\gamma}{\partial\nu}$$

$$= -\int_{\partial\omega_2} a\frac{\partial\gamma}{\partial\nu}.$$

From (38) and (52) we have

$$(63)\quad \int_{\partial\Omega} a\frac{\partial\Phi}{\partial\nu}\zeta = \int_{\partial\Omega} a\frac{\partial\zeta}{\partial\nu}\Phi.$$

Note that

$$(64)\quad \int_{\partial\Omega} a\frac{\partial\Phi}{\partial\nu}\zeta = 2\pi\zeta(\partial\omega_1) + \int_{\partial\omega_2} a\frac{\partial\Phi}{\partial\nu}h$$

and

$$(65)\quad \int_{\partial\Omega} a\frac{\partial\zeta}{\partial\nu}\Phi = \int_{\partial\omega_2} a\frac{\partial\zeta}{\partial\nu}\Phi(\partial\omega_2).$$

Combining (62), (63), (64) and (65) we find

$$\left|\int_{\partial\omega_2} a\frac{\partial\Phi}{\partial\nu}h\right| \le 2\pi|\zeta(\partial\omega_1)| - |\Phi(\partial\omega_2)|\int_{\partial\omega_2} a\frac{\partial\gamma}{\partial\nu}.$$

Using (63), (64) and (65) with $\zeta = \gamma$ we are led to

$$\Phi(\partial\omega_2)\int_{\partial\omega_2} a\frac{\partial\gamma}{\partial\nu} = 2\pi\gamma(\partial\omega_1).$$

Hence

$$(66)\quad \left|\int_{\partial\omega_2} a\frac{\partial\Phi}{\partial\nu}h\right| \le 2\pi\left(|\zeta(\partial\omega_1)| + |\gamma(\partial\omega_1)|\right) \le 4\pi$$

by (59) and (60). Since (66) holds for every h satisfying $|h| \le 1$ we obtain

$$\int_{\partial\omega_2} a\left|\frac{\partial\Phi}{\partial\nu}\right| \le 4\pi$$

which is the desired conclusion.

Lemma X.7. *Assume Φ satisfies (38)–(41) and let $1 < p < 2$. There is a constant C_p depending only on p such that*

$$(67)\quad \left(\int_\Omega |\nabla\Phi|^p\right)^{1/p} \le C_p\frac{d}{\alpha}|G|^{(1/p)-(1/2)}.$$

The proof relies on the following:

Lemma X.8. *Given a vector-field $h = (h_1, h_2)$ in Ω with $h_j \in L^q(\Omega), j = 1,2$ and $q > 2$, let $\zeta \in V$ be the unique solution of the linear problem*

(68) $$\int_\Omega a\, \nabla\zeta\, \nabla\varphi = \int_\Omega h\nabla\varphi \qquad \forall \varphi \in V$$

where

$$V = \{v \in H^1(\Omega); v = 0 \quad \text{on } \partial G \text{ and } v = \text{Const. on each } \partial\omega_i\}.$$

Then

$$\|\zeta\|_{L^\infty(\Omega)} \le \frac{C_q}{\alpha} |G|^{(1/2)-(1/q)} \|h\|_q.$$

Proof of Lemma X.7. Let q be the conjugate exponent of p. Multiplying (38) by ζ and integrating we have

(69) $$\int_\Omega a\, \nabla\Phi\, \nabla\zeta = -2\pi \sum_{i=1}^n d_i\, \zeta(\partial\omega_i).$$

On the other hand, using (68) with $\varphi = \Phi$ we see that

(70) $$\int_\Omega h\, \nabla\Phi = -2\pi \sum_{i=1}^n d_i\, \zeta(\partial\omega_i).$$

Applying Lemma X.8 we obtain

(71) $$\left| \int_\Omega h\, \nabla\Phi \right| \le 2\pi d \frac{C_q}{\alpha} |G|^{(1/2)-(1/q)} \|h\|_q.$$

Since (71) holds for every $h \in L^q$ we deduce that

$$\|\nabla\Phi\|_p \le 2\pi\, d \frac{C_q}{\alpha} |G|^{(1/2)-(1/q)}$$

which is the desired conclusion.

Proof of Lemma X.8. We follow the method of G. Stampacchia [1]. Let $k > 0$ and apply (68) with $\varphi = (\zeta - k)^+$. This yields

(72) $$\alpha \int_\Omega |\nabla(\zeta - k)^+|^2 \le \int_\Omega |h|\, |\nabla(\zeta - k)^+|.$$

Set

$$\Omega(k) = \{x \in \Omega;\ \zeta > k\}$$

From (72) we have

$$\alpha\|\nabla(\zeta-k)^+\|_2 \le \left(\int_{\Omega(k)}|h|^2\right)^{1/2} \le \|h\|_q\,\mu(k)^{(1/2)-(1/q)} \tag{73}$$

and consequently

$$\alpha\|\nabla(\zeta-k)^+\|_1 \le \|h\|_q\,\mu(k)^{1-(1/q)}. \tag{74}$$

Next, we use the Sobolev embedding $W_0^{1,1}(G)\subset L^2(G)$ to deduce that

$$\alpha S\|(\zeta-k)^+\|_2 \le \|h\|_q\,\mu(k)^{1-(1/q)} \tag{75}$$

where S is some universal constant.

Hence

$$\|(\zeta-k)^+\|_1 \le \frac{1}{\alpha S}\|h\|_q\,\mu(k)^{\gamma} \tag{76}$$

where

$$\gamma=\frac{3}{2}-\frac{1}{q}.$$

Finally, we write

$$\|(\zeta-k)^+\|_1 = -\int_k^{+\infty}(t-k)\,d\mu(t)\equiv H(k)$$

and we are led to the differential inequality

$$H'(k)=-\mu(k)\le-\left(\frac{H(k)}{\beta}\right)^{1/\gamma} \tag{77}$$

where

$$\beta=\frac{1}{\alpha S}\|h\|_q.$$

Integrating the differential inequality (77) we obtain

$$H(k)=0 \qquad \text{for } k\ge\beta^{1/\gamma}\frac{\gamma}{(\gamma-1)}H^{1-(1/\gamma)}(0). \tag{78}$$

But

$$H(0)=\|\zeta^+\|_1\le\beta|G|^{\gamma} \quad \text{by (76)}.$$

Therefore

$$H(k)=0 \quad \text{for } k\ge\beta\frac{\gamma}{(\gamma-1)}|G|^{\gamma-1},$$

i.e.,

$$\|\zeta^+\|_{L^\infty} \le \frac{1}{\alpha S}\|h\|_q \frac{\gamma}{(\gamma-1)} |G|^{(1/2)-(1/q)}$$

which completes the proof of Lemma X.8.

X.4. The basic estimates: $\int_G |\nabla v_\varepsilon|^2 \le C|\log \varepsilon|$ and $\int_G |\nabla v_\varepsilon|^p \le C_p$ for $p < 2$

We now return to our main goal, namely to estimate $|\nabla v_\varepsilon|$ in various norms, where v_ε satisfies the Ginzburg-Landau equation (1)–(2). Our main result is the following:

Theorem X.1. *Let v_ε be a solution of (1)–(2). There is a constant C depending only on g and G such that*

$$\int_G |\nabla v_\varepsilon|^2 \le C\,(|\log \varepsilon| + 1)\,. \tag{79}$$

Moreover, given any $1 < p < 2$, there is a constant C_p depending only on g, G and p such that

$$\int_G |\nabla v_\varepsilon|^p \le C_p. \tag{80}$$

The proof of Theorem X.1 relies on the following important observation: In view of (6) we have

$$\int_{B(x_i,\lambda\varepsilon)} |\nabla v_\varepsilon|^2 \le C. \tag{81}$$

Hence, it suffices to estimate $|\nabla v_\varepsilon|$ in $\Omega_\varepsilon = G \setminus \bigcup_{i\in J'} \omega_i$ where $\omega_i = B(x_i, \lambda\varepsilon)$. Here we shall use the splitting of Section X.2, namely (33) of that section:

$$|\nabla v_\varepsilon| \le 2\Big[|\nabla \Phi_\varepsilon| + |\nabla H_\varepsilon| + |\nabla \rho_\varepsilon|\Big] \qquad \text{in } \Omega_\varepsilon.$$

We will estimate each of the terms on the right-hand side of (33) separately.

Estimate for ∇H_ε:

Lemma X.9. *There is a constant C independent of ε (C depends only on g and G) such that*

$$\int_{\Omega_\varepsilon} |\nabla H_\varepsilon|^2 \leq C. \tag{82}$$

Proof. Recall that H_ε satisfies (see (27))

$$\operatorname{div}(\rho_\varepsilon^2 \nabla H_\varepsilon) = 0.$$

We claim that for $i \in L$, we have

$$\int_{\partial\omega_i} \rho_\varepsilon^2 \frac{\partial H_\varepsilon}{\partial \nu} = 0. \tag{83}$$

For simplicity we now drop the subscript ε. Recall (see (28)) that

$$\frac{\partial}{\partial x_1}(v \times v_{x_1}) + \frac{\partial}{\partial x_2}(v \times v_{x_2}) = 0 \qquad \text{in } G.$$

Integrating this relation over ω_i we obtain

$$\int_{\partial\omega_i} v \times \frac{\partial v}{\partial \nu} = 0.$$

On the other hand, by (25) and (26) we have

$$v \times \frac{\partial v}{\partial \nu} = \rho^2 \frac{\partial H}{\partial \nu} \quad \text{on } \partial\omega_i$$

(since $\dfrac{\partial \Phi}{\partial \tau} = 0$ on $\partial\omega_i$ by (16)). This proves (83).

We may now invoke Lemma X.4 to assert that

$$\operatorname*{Sup}_{\Omega} H - \operatorname*{Inf}_{\Omega} H \leq \sum_{i \in L} \left[\operatorname*{Sup}_{\partial\omega_i} H - \operatorname*{Inf}_{\partial\omega_i} H\right] + \left[\operatorname*{Sup}_{\partial\tilde{G}} H - \operatorname*{Inf}_{\partial\tilde{G}} H\right]. \tag{84}$$

For $i \in L$, we have

$$\operatorname*{Sup}_{\partial\omega_i} H - \operatorname*{Inf}_{\partial\omega_i} H \leq \int_{\partial\omega_i} \left|\frac{\partial H}{\partial \tau}\right| \leq 4\left(\int_{\partial\omega_i} \left|\frac{\partial v}{\partial \tau}\right| + \left|\frac{\partial \Phi}{\partial \nu}\right|\right). \tag{85}$$

Here we have used the fact that $\rho \geq \frac{1}{2}$ in Ω together with (25)–(26). Similarly

$$\operatorname*{Sup}_{\partial\tilde{G}} H - \operatorname*{Inf}_{\partial\tilde{G}} H \leq 4\left(\int_{\partial\tilde{G}} \left|\frac{\partial v}{\partial \tau}\right| + \left|\frac{\partial \Phi}{\partial \nu}\right|\right). \tag{86}$$

By (10) and (11), we have

$$\sum_{i\in L}\int_{\partial\omega_i}|\nabla v|\le C$$

and

$$\int_{\partial\tilde G}\left|\frac{\partial v}{\partial\tau}\right|\le C.$$

On the other hand, by Lemma X.6, we have

$$\int_{\partial\omega_j}\left|\frac{\partial\Phi}{\partial\nu}\right|\le\frac{4\pi}{\alpha}\sum_{i\in L}|\delta_i|\quad\forall j\in L,\tag{87}$$

where δ_i is given by (19), and also

$$\int_{\partial\tilde G}\left|\frac{\partial\Phi}{\partial\nu}\right|\le\frac{4\pi}{\alpha}\sum_{i\in L}|\delta_i|.\tag{88}$$

Finally we claim that

$$|\delta_i|\le C\qquad\forall i\in L,\tag{89}$$

where C is independent of ε. Indeed,

$$\delta_i=\frac{1}{2\pi}\int_{\partial\omega_i}\frac{1}{\rho^2}v\times v_\tau$$

and hence

$$|\delta_i|\le 4\int_{\partial\omega_i}|v_\tau|\le C.\tag{90}$$

Combining all the above estimates and the fact that card $L\le$ card $J\le N$ (independent of ε) we obtain

$$\operatorname*{Sup}_{\Omega}H-\operatorname*{Inf}_{\Omega}H\le C.\tag{91}$$

Set $H_0=\operatorname{Inf}_\Omega H$. We multiply (27) by $(H-H_0)$ and integrate over Ω. We obtain

$$\begin{aligned}\int_\Omega\rho^2|\nabla H|^2&=-\sum_{i\in L}\int_{\partial\omega_i}\rho^2\frac{\partial H}{\partial\nu}(H-H_0)+\int_{\partial\tilde G}\rho^2\frac{\partial H}{\partial\nu}(H-H_0)\\&\le\|H-H_0\|_{L^\infty(\Omega)}\left(\sum_{i\in L}\int_{\partial\omega_i}\rho^2\left|\frac{\partial H}{\partial\nu}\right|+\int_{\partial\tilde G}\rho^2\left|\frac{\partial H}{\partial\nu}\right|\right).\end{aligned}\tag{92}$$

By (25) and (26) we have on $\partial\omega_i$ and on $\partial\widetilde{G}$

$$\rho^2\left|\frac{\partial H}{\partial\nu}\right| \le \left|\frac{\partial v}{\partial\nu}\right| + \left|\frac{\partial\Phi}{\partial\tau}\right| = \left|\frac{\partial v}{\partial\nu}\right|.$$

Hence

$$\left(\sum_{i\in L}\int_{\partial\omega_i}\rho^2\left|\frac{\partial H}{\partial\nu}\right| + \int_{\partial\widetilde{G}}\rho^2\left|\frac{\partial H}{\partial\nu}\right|\right) \le C \tag{93}$$

by (10) and (11). Combining (91), (92) and (93) we are led to (82).

Estimates for $\nabla\Phi_\varepsilon$:

Lemma X.10. *There is a constant C depending only on g and G such that*

$$\int_\Omega |\nabla\Phi_\varepsilon|^2 \le C(|\log\varepsilon| + 1). \tag{94}$$

Proof. By Lemma X.5 we have

$$\|\Phi_\varepsilon\|_{L^\infty(\Omega)} \le 4A\,d\left(\operatorname*{Max}_{i\in L}\log\frac{|\widetilde{G}|}{|\omega_i|} + 1\right) \tag{95}$$

where $d = \sum_{i\in L}|\delta_i|$ and A is some universal constant. Note that, by (90), we have

$$d \le C \tag{96}$$

where C is independent of ε. On the other hand,

$$|\omega_i| = \pi\lambda^2\varepsilon^2$$

and therefore

$$\|\Phi_\varepsilon\|_{L^\infty(\Omega)} \le C\,(|\log\varepsilon| + 1) \tag{97}$$

where C depends only on g and G.

Next we multiply (15) by Φ_ε. We obtain

$$\int_\Omega \frac{1}{\rho_\varepsilon^2}|\nabla\Phi_\varepsilon|^2 = -\sum_{i\in L}\int_{\partial\omega_i}\frac{1}{\rho_\varepsilon^2}\frac{\partial\Phi_\varepsilon}{\partial\nu}\,\Phi_\varepsilon = -\sum_{i\in L}2\pi\delta_i\,\Phi_\varepsilon(\partial\omega_i)$$
$$\le 2\pi\|\Phi_\varepsilon\|_{L^\infty(\Omega)}d \le C(|\log\varepsilon| + 1),$$

by (97).

Lemma X.11. *Given any $1 < p < 2$, there is a constant C_p depending only on g, G and p such that*

$$\int_{\Omega_\varepsilon} |\nabla \Phi_\varepsilon|^p \le C_p. \tag{98}$$

Proof. Estimate (98) is a direct consequence of Lemma X.7 and (96).

Estimates for $\nabla \rho_\varepsilon$:

Lemma X.12. *There is a constant C depending only on g and G such that*

$$\int_{\Omega} |\nabla \rho_\varepsilon|^2 \le C(|\log \varepsilon| + 1). \tag{99}$$

Proof. Using (31) we rewrite (14) as

$$-\Delta \rho + \frac{1}{\rho^3}|v \times \nabla v|^2 = \frac{1}{\varepsilon^2}\rho(1-\rho^2). \tag{100}$$

We multiply (100) by $(\rho - 1)$ and integrate over Ω:

$$\begin{aligned}
\int_{\Omega} |\nabla \rho|^2 &= \int_{\partial \tilde{G}} \frac{\partial \rho}{\partial \nu}(\rho - 1) - \sum_{i \in L} \int_{\partial \omega_i} \frac{\partial \rho}{\partial \nu}(\rho - 1) \\
&\quad + \int_{\Omega} \frac{(1-\rho)}{\rho^3}|v \times \nabla v|^2 - \frac{1}{\varepsilon^2}\int_{\Omega} \rho(1-\rho)^2(1+\rho) \\
&\le \int_{\partial \Omega} |\nabla v| + 8 \int_{\Omega} |v \times \nabla v|^2 \\
&\le C + 16 \int_{\Omega} (|\nabla \Phi|^2 + |\nabla H|^2) \qquad \text{by (10), (11) and (29).}
\end{aligned}$$

The conclusion follows from Lemma X.9 and Lemma X.10.

Lemma X.13. *Given any $1 < p < 2$ there are constants α and C depending only on g, G and p such that*

$$\int_{\Omega_\varepsilon} |\nabla \rho_\varepsilon|^p \le C \varepsilon^\alpha. \tag{101}$$

Proof. We introduce the set

$$S = \{x \in \Omega\,;\, \rho > 1 - \varepsilon^\beta\} \quad \text{for some } \beta \in (0,1)$$

and

$$\overline{\rho} = \operatorname{Max}\{\rho, 1 - \varepsilon^\beta\}$$

so that $\rho = \overline{\rho}$ on S. We multiply (100) by $(1-\overline{\rho})$ and integrate over Ω:

$$-\int_\Omega \nabla\rho\cdot\nabla\overline{\rho} + \int_\Omega \frac{(1-\overline{\rho})}{\rho^3}|v\times\nabla v|^2 \tag{102}$$

$$= \int_\Omega \rho\frac{(1-\rho^2)(1-\overline{\rho})}{\varepsilon^2} - \sum_{i\in L}\int_{\partial\omega_i}\frac{\partial\rho}{\partial\nu}(1-\overline{\rho}) + \int_{\partial\tilde{G}}\frac{\partial\rho}{\partial\nu}(1-\overline{\rho}).$$

Note that

$$(1-\overline{\rho}) \le \varepsilon^\beta,$$

so that (102) yields

$$\int_S |\nabla\rho|^2 \le 8\varepsilon^\beta\left(\int_\Omega |v\times\nabla v|^2 + C\right)$$

(here we have used (10) and (11)). As in the proof of Lemma X.12 we have

$$\int_\Omega |v\times\nabla v|^2 \le 2\left(\int_\Omega |\nabla\Phi|^2 + |\nabla H|^2\right) \le C|\log\varepsilon|$$

and hence

$$\int_S |\nabla\rho|^2 \le C\varepsilon^\beta\left(|\log\varepsilon|+1\right) \tag{103}$$

where C depends only on g and G. On $\Omega\setminus S$, we write

$$\begin{aligned}\int_{\Omega\setminus S}|\nabla\rho|^p &\le \left(\int_\Omega |\nabla\rho|^2\right)^{p/2}|\Omega\setminus S|^{1-(p/2)}\\ &\le C\left(|\log\varepsilon|+1\right)^{p/2}|\Omega\setminus S|^{1-(p/2)} \quad \text{by (99).}\end{aligned} \tag{104}$$

Since

$$\frac{1}{\varepsilon^2}\int_G (1-\rho)^2 \le C$$

we deduce that

$$|\Omega\setminus S| \le C\,\varepsilon^{2-2\beta}. \tag{105}$$

Combining (103), (104) and (105) we obtain (101).

Proof of Theorem X.1. All the conclusions in the theorem follow from (81), (33) and Lemmas X.9, X.10, X.11, X.12 and X.13.

X.5. v_{ε_n} converges to $v_\star$

We may extract a subsequence $\varepsilon_n \to 0$ such that

$$\text{card } J'_{\varepsilon_n} \equiv \text{Const.} = N_1 \tag{106}$$

$$x_i^{\varepsilon_n} \to \ell_i \in \overline{G} \quad \forall i \in J'. \tag{107}$$

We cannot exclude the possibility that some of the points $(x_i^{\varepsilon_n})$ converge to the same limit. We denote by

$$a_1, a_2, \dots, a_{N_0} \qquad \text{with } N_0 \leq N_1$$

the collection of distinct points in ℓ_i.

Using Theorem X.1, we may extract a further subsequence such that

$$v_{\varepsilon_n} \to v_\star \text{ in } W^{1,p}(G) \text{ weakly, for } 1 < p < 2. \tag{108}$$

Since by (4) $|v_{\varepsilon_n}| \to 1$ in L^2, we obtain

$$v_\star \in W_g^{1,r}(G; S^1) \qquad \text{for } 1 < r < 2.$$

Passing to the limit in (28), we find

$$\frac{\partial}{\partial x_1}(v_\star \times v_{\star x_1}) + \frac{\partial}{\partial x_2}(v_\star \times v_{\star x_2}) = 0. \tag{109}$$

Note that Φ_ε and H_ε are only defined in Ω_ε. Therefore, we extend Φ_ε in G by setting

$$\Phi_\varepsilon = C_i \quad \text{in} \quad \omega_i, \ \forall i \in L,$$
$$\Phi_\varepsilon = 0 \quad \text{in} \quad G \setminus \widetilde{G}_\varepsilon.$$

We extend H_ε by its harmonic extension in ω_i if $i \in L$, and by the solution $\widetilde{H}_\varepsilon$ of

$$\Delta \widetilde{H}_\varepsilon = 0 \qquad \text{in } \omega_i \cap G,$$
$$\widetilde{H}_\varepsilon = H_\varepsilon \quad \text{on } \partial\omega_i \cap G,$$
$$\frac{\partial \widetilde{H}_\varepsilon}{\partial \nu} = 0 \qquad \text{on } \partial G \cap \omega_i,$$

if $i \in K$. We still denote by Φ_ε and H_ε the extended functions.

Clearly, by (17), we have

$$\Phi_\varepsilon = 0 \qquad \text{on } \partial G,$$

and by Lemma X.11

$$\int_G |\nabla \Phi_\varepsilon|^p \leq C_p \quad \forall\, 1 < p < 2. \tag{110}$$

Since $B(x_i, 2\lambda\varepsilon) \cap B(x_j, 2\lambda\varepsilon) = \emptyset$ if $i \neq j$, the distances between ω_i and ω_j are larger than $2\lambda\varepsilon$, and therefore by the trace theorem together with Lemma X.9, and the definition of H_ε we see (as in Lemma 3 of H. Brezis, F. Merle and T. Rivière [1]) that

$$\int_{\omega_i} |\nabla H_\varepsilon|^2 \leq C \qquad \forall i \in J',$$

where C depends only on g and G. Combining this inequality with (82) we obtain

$$\int_G |\nabla H_\varepsilon|^2 \leq C. \tag{111}$$

Since H_ε is defined up to a constant, we may impose the condition

$$\int_G H_\varepsilon = 0. \tag{112}$$

In view of (110), (111) and (112) we may extract a further subsequence $\varepsilon_n \to 0$ such that

$$\Phi_{\varepsilon_n} \to \Phi_\star \quad \text{weakly in } W^{1,p},\ 1 < p < 2, \tag{113}$$

$$H_{\varepsilon_n} \to H_\star \quad \text{weakly in } H^1, \tag{114}$$

where $\Phi_\star \in W_0^{1,p}(G;\mathbb{R})$ and $H_\star \in H^1(\Omega;\mathbb{R})$ satisfies

$$\int_G H_\star = 0. \tag{115}$$

Since $\rho_\varepsilon \to 1$, we may pass to the limit in (15) and (27), so that

$$\Delta \Phi_\star = 0 \quad \text{in } G \setminus \cup\{a_i\}, \tag{116}$$

(117) $$\Delta H_\star = 0 \quad \text{in } G \setminus \cup\{a_i\}.$$

Since H belongs to $H^1(G)$, (117) yields

(118) $$\Delta H_\star = 0 \quad \text{in } G.$$

Therefore, $\Phi_\star$ is smooth in $G \setminus \bigcup\{a_i\}$ and $H_\star$ is smooth in G. On the other hand, passing to the limit in (25) and (26) we obtain

(119) $$v_\star \times v_{\star x_1} + \Phi_{\star x_2} = H_{\star x_1},$$

(120) $$v_\star \times v_{\star x_2} - \Phi_{\star x_1} = H_{\star x_2}.$$

Hence

(121) $$v_\star \in C^\infty\left(G \setminus \cup\{a_i\}; S^1\right).$$

Moreover we deduce from (109) that $v_\star$ is a harmonic S^1-valued map in $G \setminus \bigcup\{a_i\}$, i.e.,

(122) $$-\Delta v_\star = v_\star |\nabla v_\star|^2 \quad \text{in } G \setminus \cup\{a_i\}.$$

Theorem X.2. *We have, for any compact subset K of $G \setminus \bigcup\{a_i\}$,*

(123) $$v_{\varepsilon_n} \to v_\star \quad \text{in } C^k(K) \quad \forall k \in \mathbb{N}$$

(124) $$\frac{1 - |v_{\varepsilon_n}|^2}{\varepsilon_n^2} \to |\nabla v_\star|^2 \quad \text{in } C^k(K) \text{ as } n \to +\infty.$$

Proof. The proof is divided into three steps.

Step 1 : We have, for any compact subset K of $G \setminus \bigcup\{a_i\}$,

(125) $$\Phi_{\varepsilon_n} \to \Phi_\star \quad \text{strongly in } H^1(K),$$

(126) $$H_{\varepsilon_n} \to H_\star \quad \text{strongly in } H^1(K),$$

and

(127) $$\rho_{\varepsilon_n} \to 1 \quad \text{strongly in } H^1(K).$$

Proof. Let ζ be a smooth function compactly supported in $G \setminus \bigcup\{a_i\}$ such that

$$\zeta \equiv 1 \quad \text{on } K.$$

For n sufficiently large, the support of ζ is in Ω_{ε_n} and therefore we may multiply (15) by $\zeta\,(\Phi_{\varepsilon_n} - \Phi_\star)$ and integrate on G. We obtain

$$\int_G \frac{\zeta}{\rho_{\varepsilon_n}^2}|\nabla\Phi_{\varepsilon_n}|^2 + \frac{1}{\rho_{\varepsilon_n}^2}(\Phi_{\varepsilon_n} - \Phi_\star)\nabla\Phi_{\varepsilon_n}\nabla\zeta = \int_G \frac{\zeta}{\rho_{\varepsilon_n}^2}\nabla\Phi_{\varepsilon_n}\nabla\Phi_\star\,. \tag{128}$$

From (110) and the Sobolev embedding theorem we deduce that

$$\|\Phi_{\varepsilon_n} - \Phi_\star\|_{L^q} \to 0 \quad \forall\ q < +\infty, \qquad \text{as } n \to +\infty \tag{129}$$

and hence

$$\int \frac{1}{\rho_{\varepsilon_n}^2}(\Phi_{\varepsilon_n} - \Phi_\star)\nabla\Phi_\varepsilon\nabla\zeta \to 0 \qquad n \to +\infty. \tag{130}$$

Since Φ_{ε_n} converges weakly to $\Phi_\star$ in $W^{1,p}$ and since $\rho_\varepsilon \to 1$, we obtain

$$\int_G \frac{\zeta}{\rho_{\varepsilon_n}^2}\nabla\Phi_{\varepsilon_n}\nabla\Phi_\star \to \int_G \zeta|\nabla\Phi_\star|^2 \quad \text{as } n \to +\infty. \tag{131}$$

Hence, combining (128), (130) and (131) we have

$$\int_G \frac{1}{\rho_{\varepsilon_n}^2}\zeta|\nabla\Phi_{\varepsilon_n}|^2 \to \int_G \zeta|\nabla\Phi_\star|^2 \quad \text{as } n \to +\infty.$$

Since $\rho_{\varepsilon_n} \le 1$ it follows that

$$\int_G \zeta|\nabla\Phi_{\varepsilon_n}|^2 \le \int_G \zeta|\nabla\Phi_\star|^2 + o(1)\,.$$

By a standard lower-semicontinuity argument we deduce that

$$\nabla\Phi_{\varepsilon_n} \to \nabla\Phi_\star \quad \text{strongly in } L^2(K).$$

Similarly, using equation (27), we derive that

$$\int_G \rho_{\varepsilon_n}^2\,\zeta|\nabla H_{\varepsilon_n}|^2 \to \int_G \zeta|\nabla H_\star|^2 \quad \text{as } n \to +\infty. \tag{132}$$

We now prove (126). Indeed we have

$$\int_G \rho_{\varepsilon_n}^2 \zeta |\nabla(H_{\varepsilon_n} - H_\star)|^2 = \int_G \rho_{\varepsilon_n}^2 \zeta |\nabla H_{\varepsilon_n}|^2 \\ -2\int_G \rho_{\varepsilon_n}^2 \zeta \nabla H_{\varepsilon_n} \nabla H_\star + \int_G \rho_{\varepsilon_n}^2 \zeta |\nabla H_\star|^2. \tag{133}$$

Note that

$$\int_G \rho_{\varepsilon_n}^2 \zeta \nabla H_{\varepsilon_n} \nabla H_\star \to \int_G \zeta |\nabla H_\star|^2 \tag{134}$$

and

$$\int_G \rho_{\varepsilon_n}^2 \zeta |\nabla H_\star|^2 \to \int_G \zeta |\nabla H_\star|^2 . \tag{135}$$

Combining (132), (133), (134) and (135) we obtain (126).

We now turn to (127). Multiplying (14) by $\zeta(1-\rho_\varepsilon)$ and using (31) we obtain

$$\int_G \zeta |\nabla \rho_\varepsilon|^2 - \int_G (1-\rho_\varepsilon)\nabla \rho_\varepsilon \nabla \zeta = \int_\Omega \zeta \frac{(1-\rho_\varepsilon)}{\rho_\varepsilon^3} |v_\varepsilon \times \nabla v_\varepsilon|^2 \\ - \frac{1}{\varepsilon^2}\int_G \zeta \rho_\varepsilon (1-\rho_\varepsilon)^2 (1+\rho_\varepsilon). \tag{136}$$

Since $\rho_\varepsilon \to 1$ in $W^{1,p}$, we are led (applying (29)) to

$$\int_G \zeta |\nabla \rho_\varepsilon|^2 \leq 16 \int_G \zeta (1-\rho_\varepsilon)(|\nabla H_\varepsilon|^2 + |\nabla \Phi_\varepsilon|^2) + o(1). \tag{137}$$

Using (125), (126), the fact that $\rho_\varepsilon \to 1$ a.e. and Lebesgue's dominated convergence theorem, we see that the right-hand side of (137) tends to zero as $n \to +\infty$. Hence

$$\int_G \zeta |\nabla \rho_{\varepsilon_n}|^2 \to 0 \quad \text{as } n \to +\infty.$$

which yields (127).

Step 2: For any compact subset K of $G \setminus \bigcup\{a_i\}$ we have

$$v_{\varepsilon_n} \to v_\star \qquad \text{in } H^1(K). \tag{138}$$

Proof. Using Step 1, (25)–(26) and (119)–(120) we have

$$v_{\varepsilon_n} \times \nabla v_{\varepsilon_n} \to v_\star \times \nabla v_\star \quad \text{in } L^2(K). \tag{139}$$

On K we may write locally

$$v_{\varepsilon_n} = \rho_{\varepsilon_n} e^{i\psi_{\varepsilon_n}} \quad \text{and } v_\star = e^{i\psi_\star} \tag{140}$$

so that

$$v_{\varepsilon_n} \times \nabla v_{\varepsilon_n} = \rho_{\varepsilon_n}^2 \nabla \psi_{\varepsilon_n} \quad \text{and } v_\star \times \nabla v_\star = \nabla \psi_\star.$$

Hence by (139) and (127) we obtain

$$\nabla \psi_{\varepsilon_n} \to \nabla \psi_\star \quad \text{in } L^2(K) \tag{141}$$

and (138) follows from (140), (141) and (127).

Step 3: Proof of (123) and (124)

They are direct consequences of (138) and the methods developed in F. Bethuel, H. Brezis and F. Hélein [2].

Theorem X.3. *We have for any compact subset K of $\overline{G} \setminus \bigcup\{a_i\}$*

$$v_{\varepsilon_n} \to v_\star \quad \textit{in } C^{1,\alpha}(K), \quad \forall\, 0 < \alpha < 1 \tag{142}$$

$$\left\| \frac{1 - |v_{\varepsilon_n}|^2}{\varepsilon_n^2} \right\|_{L^\infty(K)} \le C_K. \tag{143}$$

Proof. Let $x_0 \in \partial G \setminus \bigcup\{a_i\}$ and let $R > 0$ be such that

$$B(x_0, 2R) \cap (\cup\{a_i\}) = \emptyset.$$

We are going to prove that

$$v_{\varepsilon_n} \to v_\star \quad \text{in } H^1(B(x_0, R) \cap G). \tag{144}$$

Let ζ be a smooth function in $\mathbb{R}^2$ such that

$$\zeta \equiv 1 \quad \text{in } B(x_0, R) \tag{145}$$

$$\zeta \equiv 0 \quad \text{in } \mathbb{R}^2 \setminus B(x_0, 2R). \tag{146}$$

Multiplying (15) by $\zeta \Phi_{\varepsilon_n}$ and (116) by $\zeta \Phi_\star$ we obtain (using the fact that $\Phi_{\varepsilon_n} = \Phi_\star = 0$ on $\partial G \cap B(x_0, 2R)$) as in Step 1 of the proof of Theorem X.2

$$\int_G \zeta |\nabla \Phi_{\varepsilon_n}|^2 \le \int \zeta |\nabla \Phi_\star|^2 \ + o(1)$$

and hence

(147) $$\Phi_{\varepsilon_n} \to \Phi_\star \quad \text{in } H^1(B(x_0,R)\cap G).$$

For H_{ε_n} recall that

$$\frac{\partial H_{\varepsilon_n}}{\partial \nu} = v_{\varepsilon_n} \times \frac{\partial v_{\varepsilon_n}}{\partial \nu} \quad \text{on } \partial\widetilde{G}_\varepsilon$$

and hence, by (4),

(148) $$\int_{\partial G} \left|\frac{\partial H_\varepsilon}{\partial \nu}\right|^2 \le C.$$

Multiplying (27) by $\zeta(H_\varepsilon - H_\star)$ and integrating, we obtain, after computations similar to those in Step 1 of the proof of Theorem X.2

(149) $$\int_G \rho_{\varepsilon_n}^2 \zeta |\nabla H_{\varepsilon_n}|^2 = \int_G \zeta |\nabla H|^2 + \int_{\partial G} \zeta \frac{\partial H_{\varepsilon_n}}{\partial \nu}(H_{\varepsilon_n} - H_\star) + o(1).$$

Since $H_{\varepsilon_n} \to H_\star$ weakly in $H^1(G)$, $H_{\varepsilon_n} \to H_\star$ weakly in $H^{1/2}(\partial G)$ and therefore by the Sobolev embedding

(150) $$H_{\varepsilon_n} \to H_\star \quad \text{strongly in } L^2(\partial G).$$

Combining (148) and (150), we have

$$\left|\int_{\partial G} \zeta \frac{\partial H_{\varepsilon_n}}{\partial \nu}(H_{\varepsilon_n} - H_\star)\right| \to 0 \quad \text{as } n \to +\infty.$$

Hence, we obtain

(151) $$\int_G \rho_{\varepsilon_n}^2 \zeta |\nabla H_{\varepsilon_n}|^2 = \int_G \zeta |\nabla H|^2 + o(1).$$

Arguing as in Step 1 of the proof of Theorem X.2, we deduce from (151) that

(152) $$\nabla H_{\varepsilon_n} \to \nabla H_\star \quad \text{strongly in } L^2(B(x_0,R)\cap G).$$

Similarly, arguing as in Step 1 (using the fact that $\rho_\varepsilon - 1 = 0$ on ∂G), we may prove that

(153) $$\rho_\varepsilon \to 1 \quad \text{in } H^1(B(x_0,R)\cap G).$$

Combining (147), (152) and (153) we prove (144) as in Step 3 of the proof of Theorem X.2. Finally (142) and (143) follow from boundary estimate techniques developed in F. Bethuel, H. Brezis and F. Hélein [2].

Remark X.1. We emphasize that Theorems X.1, X.2 and X.3 are proved under the assumption that G is starshaped. We do not know whether the conclusion still holds for a general simply connected domain. Note, however, that the conclusion may fail when Ω is not simply connected. Here is a simple example. Take $G = B_2 \setminus \overline{\Omega}$ where Ω is simply connected but not convex and $\Omega \subset B_1$. Let

$$g = \begin{cases} +1 & \text{on } \partial B_2 \\ -1 & \text{on } \partial\Omega. \end{cases}$$

Let v_ε be a real-valued solution of

$$-\Delta v_\varepsilon = \frac{1}{\varepsilon^2} v_\varepsilon(1 - |v_\varepsilon|^2) \quad \text{in } G,$$
$$v_\varepsilon = g \quad \text{on } \partial G$$

obtained by minimizing the Ginzburg-Landau energy E_ε in $H^1_g(G;\mathbb{R})$. One can prove (using the same techniques as in R. Kohn and P. Sternberg [1]) that

$$v_\varepsilon \to v_\star = \begin{cases} +1 & \text{in } G \setminus \overline{\text{conv}\,\Omega} \\ -1 & \text{in } \text{Int}(\text{conv}\,\Omega \setminus \overline{\Omega}) \end{cases}$$

where conv denotes the convex hull.

X.6. Properties of $v_\star$

We have already proved in Section X.5 that $v_\star$ is a smooth harmonic map from $G \setminus \bigcup\{a_i\}$ into S^1. We are first going to show that all the points (a_i) lie in G and not on the boundary ∂G.

Theorem X.4. *We have*

(154) $$a_i \in G \quad \forall i = 1, 2, \ldots, N_0$$

and in particular $v_\star$ is smooth near the boundary ∂G.

The proof relies on the following:

Lemma X.14. *Let $b_1, b_2, \ldots, b_\ell$ be ℓ points in $\overline{G}$ and let v be a smooth harmonic map from $\overline{G} \setminus \bigcup\{b_i\}$ into S^1 such that*

(155) $$v \in W^{1,p}(G) \quad \textit{for every } p \in (1,2),$$

(156) $$v = g \qquad \textit{on } \partial G \textit{ and } g \textit{ is smooth},$$

(157) $$\int_{\partial G} \left|\frac{\partial v}{\partial \nu}\right|^2 < \infty.$$

Then v is smooth in some neighborhood of the boundary ∂G.

Proof of Lemma X.14. Assume that one of the points (b_i), say b_1, lies on ∂G. We are going to prove that v is smooth in $G \cap B(b_1, R)$ for some R. For simplicity take $b_1 = 0$.

Choose $R > 0$ so small that $B(0, R)$ contains no other singularity other than 0 and that $G \cap B(0, R)$ is simply connected. In $G \cap B(0, R)$ we may write

$$v = e^{i\varphi} \tag{158}$$

where φ is a real-valued smooth harmonic function in $\overline{G} \cap (B(0, R) \setminus \{0\})$.

From (155) we deduce that $\varphi \in W^{1,p}(G \cap B(0, R))$. On $\partial G \cap B(0, R)$ we may write

$$g = e^{i\varphi_0} \tag{159}$$

where φ_0 is some smooth real-valued function on $\partial G \cap B(0, R)$.

We may assume that the tangent vector to ∂G at 0 is along the x_1-axis and we set

$$\Gamma_+ = \{x \in \partial G \cap B(0, R);\ x_1 > 0\}$$
$$\Gamma_- = \{x \in \partial G \cap B(0, R);\ x_1 < 0\}.$$

Choosing R still smaller if necessary we may assume that Γ_+ and Γ_- are connected. Combining (156), (158), (159) and the fact that v is smooth on $\Gamma_+ \cup \Gamma_-$ we deduce that there are two integers k_+ and k_- such that, near 0,

$$\begin{cases} \varphi = \varphi_0 + 2\pi k_+ & \text{on } \Gamma_+, \\ \varphi = \varphi_0 + 2\pi k_- & \text{on } \Gamma_-. \end{cases} \tag{160}$$

We claim that

$$k_+ = k_-. \tag{161}$$

Proof of (161). From (157) we deduce that

$$\int_{\partial G \cap B(0,R)} \left|\frac{\partial \varphi}{\partial \nu}\right|^2 < \infty. \tag{162}$$

Since φ is harmonic in $G \cap B(0,R)$ it follows that $\dfrac{\partial \varphi}{\partial \nu}$ makes sense as a distribution on $\partial G \cap B(0,R)$ (see e.g., J. L. Lions and E. Magenes [1]). In view of (162) and a celebrated result of L. Schwartz [1] we may write

$$\frac{\partial \varphi}{\partial \nu} = h + \sum_{\text{finite}} c_\alpha D^\alpha \delta \quad \text{in } \mathcal{D}'(\partial G \cap B(0,R))$$

where $h \in L^2(\partial G \cap B(0,R))$ and D denotes tangential derivation.

Since the fundamental solution for the Neumann problem has a logarithmic behavior it follows that

$$\varphi = \frac{1}{2\pi} \sum c_\alpha D^\alpha \log(|x|^{-1}) + \varphi_1 + \text{a smooth function}$$

where φ_1 satisfies

$$\begin{aligned} \Delta \varphi_1 &= 0 \quad \text{in } G \cap B(0,R), \\ \frac{\partial \varphi_1}{\partial \nu} &= h \quad \text{on } \partial G \cap B(0,R). \end{aligned}$$

Using a Pohozaev-type identity one deduces that $\varphi_1 \in H^1(\partial G \cap B(0,R))$. Combining this fact with (160) we see that $c_\alpha = 0$, $\forall \alpha$ and that $k_+ = k_-$.

Proof of Lemma X.14 completed. By adding a constant to φ_0 and using (161) we may now assume that

$$\varphi = \varphi_0 \qquad \text{on } \partial G \cap B(0,R).$$

Since φ is harmonic in $G \cap B(0,R)$ we deduce that φ is smooth in $\overline{G} \cap B(0,R)$ and thus v is also smooth in $\overline{G} \cap B(0,R)$.

Proof of Theorem X.4. Recall (see (4)) that

$$\int_{\partial G} \left| \frac{\partial v_{\varepsilon_n}}{\partial \nu} \right|^2 \leq C. \tag{163}$$

Passing to the limit in (163) with the help of Theorem X.3 we deduce that

$$\int_{\partial G} \left| \frac{\partial v_\star}{\partial \nu} \right|^2 < \infty.$$

All the conditions of Lemma X.14 are satisfied and we may thus assert that $v_\star$ is smooth in some neighborhood of the boundary ∂G.

We now turn to the proof of (154). Assume, by contradiction, that one of the points (a_i), say a_1 belongs to ∂G. From the definition of the points (a_i) we see that given any $r > 0$, $B(a_1, r)$ contains at least one bad disc $B(x_i^{\varepsilon_n}, 2\lambda\varepsilon_n)$ for all $n \geq N(r)$ and therefore

$$\frac{1}{\varepsilon^2}\int_{B(a_1,r)\cap G}(|v_{\varepsilon_n}|^2-1)^2 \geq \mu_0 \quad \forall n \geq N(r). \tag{164}$$

On the other hand, if we multiply equation (1) by $\sum_{i=1}^{2}(x_i - a_1)\frac{\partial v_\varepsilon}{\partial x_i}$ and integrate over $D = B(a_1, r) \cap G$ we find (as in (35) of Chapter VII)

$$\begin{aligned}\frac{1}{2\varepsilon^2}\int_{B(a_1,r)\cap G}(|v_\varepsilon|^2-1)^2 \leq 2r\int_{\partial D}|\nabla v_{\varepsilon_n}|^2 \\ +\frac{r}{4\varepsilon^2}\int_{\partial B(a_1,r)\cap G}(|v_\varepsilon|^2-1)^2.\end{aligned} \tag{165}$$

Passing to the limit in (165) and using (164) we are led to

$$\frac{\mu_0}{2} \leq 2r(\int_{G\cap\partial B(a_1,r)}|\nabla v_\star|^2 + C) \quad \forall r < R \tag{166}$$

where C is independent of r.

Here we have used the fact that

$$\partial D = [G \cap \partial B(a_1, r)] \cup [\partial G \cap B(a_1, r)]$$

combined with Theorem X.3 and the estimate (163). We are led to a contradiction since r in (166) is arbitrarily small.

Theorem X.5. *The map $v_\star$ is the canonical harmonic map associated to (a_i, d_i), that is,*

$$v_\star(z) = \left(\frac{z-a_1}{|z-a_1|}\right)^{d_1}\cdots\left(\frac{z-a_{N_0}}{|z-a_{N_0}|}\right)^{d_{N_0}} e^{i\varphi(z)} \tag{167}$$

where φ is a smooth harmonic function in G.

Moreover the configuration (a_i, d_i) is critical for the renormalized energy W.

In addition

$$\frac{1}{4\varepsilon_n^2}(1-|v_{\varepsilon_n}|^2)^2 \to \frac{\pi}{2}\sum_i d_i^2\delta_{a_i} \tag{168}$$

in the sense of measures on $\overline{G}$.

Proof. Since $v_\star$ is in $W^{1,p}(G)$, for $1 \leq p < 2$ and satisfies (109), the analysis of Chapter I shows that $v_\star$ is the canonical harmonic map associated to the singularities (a_i, d_i), and thus (167) holds (see Corollary I.2).

Next, we note that the results in Chapter VII can be carried over to our situation, namely with u_{ε_n} replaced by v_{ε_n}. In particular, this yields (168) and the fact that

$$\nabla\varphi(a_i) = 0 \qquad \text{for } i = 1, \dots, N_0. \tag{169}$$

From Chapter VIII, we know that (169) is equivalent to the fact that (a_i, d_i) is critical for W.

CHAPTER XI

Open problems

Problem 1. Assume G is simply connected but not starshaped. Do the conclusions of Theorems 0.1, 0.2, 0.3, 0.5 and 0.6 still remain valid?

[As we have emphasized in Section III.3 our approach relies heavily on the estimate $\int_G (|u_\varepsilon|^2 - 1)^2 \leq C\varepsilon^2$ which is proved by a Pohozaev-type argument using the assumption that G is starshaped].[1]

A related question (even for starshaped domains) is the following:

Problem 2. Consider an energy of the form

$$E_\varepsilon(u) = \frac{1}{2}\int_G |\nabla u|^2 + \frac{1}{4\varepsilon^2}\int_G (|u|^2 - 1)^2 w(x)$$

where $w(x)$ is a given smooth function in $\overline{G}$ such that $w > 0$ in $\overline{G}$. Do the conclusions of Theorems 0.1, 0.2, 0.3 and 0.5 still remain valid?

Problem 3. Assume G is connected but not simply connected—for example, an annulus. What happens to the conclusions of Theorems 0.1, 0.2, 0.3 and 0.5?

[In physical experiments one often works in a 3-dimensional domain bounded by two coaxial circular cylinders; its cross-section is an annulus.]

Problem 4. Let $G = \{(x_1, x_2);\ (x_1 - 1)^2 + x_2^2 < R^2\}$ with $R < 1$. Replace equation (3) of the Introduction by

$$-u_{x_1x_1} - \frac{1}{x_1}u_{x_1} - u_{x_2x_2} = \frac{1}{\varepsilon^2}u(1 - |u|^2) \quad \text{in } G. \tag{1}$$

Study the corresponding minimization problem and its limit as $\varepsilon \to 0$.

[This comes up naturally when dealing with the cross-section of a 3-dimensional solid torus having axial symmetry.]

Problem 5. Assume G is starshaped, for example a disc. Is it possible to construct a boundary condition g and a sequence v_{ε_n} of (non-minimizing)

---

[1] After our work was completed a partial answer to Problem 1 was given by M. Struwe [1], [2].

solutions of the Ginzburg-Landau equation (3) in the Introduction such that $v_{\varepsilon_n} \to v_\star$ and $v_\star$ has singularities both of positive and negative degrees? In particular is it possible to construct g and $v_{\varepsilon_n} \to v_\star$ such that $\deg(g, \partial G) = 0$ and $v_\star$ has two singularities of degrees $+1$ and -1?

Problem 6. From Theorems 0.1 and 0.2 we know that $u_{\varepsilon_n} \to u_\star$ having singularities (a_i) and that the configuration (a_i) minimizes W. Conversely, given a configuration (a_i), which is a nondegenerate minimizer of W, is there a sequence (u_{ε_n}) of minimizers for E_{ε_n} such that $u_{\varepsilon_n} \to u_\star$ having (a_i) as its singular set?

Similarly, given a configuration (a_i, d_i), which is a nondegenerate critical point of W, is there a sequence v_{ε_n} of solutions of the Ginzburg-Landau equation (3) in the Introduction such that $v_{\varepsilon_n} \to v_\star$ having (a_i, d_i) as its singular set?

Problem 7. Let u_ε be a minimizer of E_ε as in Theorem 0.1. Prove (or disprove) that various quantities remain bounded as $\varepsilon \to 0$:

(i) $A_\varepsilon = \int_G (1 - |u_\varepsilon|)^\alpha |\nabla u_\varepsilon|^2$, for any $\alpha > 0$,

(ii) $B_\varepsilon = \int_G (1 - |u_\varepsilon|)^\alpha |u_\varepsilon|^\alpha \left|\nabla \left(u_\varepsilon/|u_\varepsilon|\right)\right|^2$, for any $\alpha > 0$,

(iii) $C_\varepsilon = \int_G |\det(\nabla u_\varepsilon)|$.

Problem 8. Study the weak solutions $u \in W^{1,1}(G; S^1)$ of the equation

$$\frac{\partial}{\partial x_1}(u \times u_{x_1}) + \frac{\partial}{\partial x_2}(u \times u_{x_2}) = 0. \tag{2}$$

In particular, is u smooth except at a finite number of points? Or else, what can be said about its singular set Σ?

[Note that if u is a smooth map from G into S^1 then (2) holds if and only if u is a harmonic map].[2]

Problem 9. For each real $p \in (1, 2)$ consider the minimization problem

$$\operatorname*{Min}_{W_g^{1,p}(G;S^1)} \int_G |\nabla u|^p. \tag{3}$$

Note that $W_g^{1,p}(G; S^1) \neq \emptyset$ even if $\deg(g, \partial G) \neq 0$. One knows (see B. Chen and R. Hardt [1]) that every minimizer u_p of (3) has only a finite number of singularities having degree $+1$ or -1. Does $\lim_{p_n \to 2} u_{p_n}$ exist? Does it

[2] After our work was completed, L. Almeida [1], [2] has constructed a solution to (2) everywhere discontinuous.

have the same properties as $u_\star = \lim\, u_{\varepsilon_n}$ stated in Theorems 0.1, 0.2, 0.3, 0.4 and 0.5?[3]

Problem 10. Assume $G = B_1$ is the unit disc and let $g(\theta) = e^{i\theta}$ on ∂G. Let u_ε be a minimizer for E_ε. Can one say that, for every $\varepsilon > 0$, u_ε has the form

$$u_\varepsilon(r,\theta) = e^{i\theta} f_\varepsilon(r)\,? \tag{4}$$

[We already know that (4) holds for ε **large**; see Theorem VIII.7 in Section VIII.5.]

Same question if we assume only that u_ε is a solution of the Ginzburg-Landau equation (3) in the Introduction.

Problem 11. Assume $G = B_1$ and $g(\theta) = e^{2i\theta}$. Let u_ε be a minimizer for (0.2). We know that for ε **large** (see Section VIII.3) u_ε has only one zero (namely, $x = 0$) and for ε **small** u_ε has precisely **two** zeroes (see Section IX.1). Is there some critical value ε_2 such that for $\varepsilon \geq \varepsilon_2$, u_ε has one zero, and for $\varepsilon < \varepsilon_2$, u_ε has two zeroes? Can one study this problem via a **bifurcation** analysis? Same question for $g(\theta) = e^{di\theta}$, with u_ε having one zero when ε is large and d zeroes when ε is small. What is the dividing line ε_d? How does ε_d depend on d? Is this phenomenon related to the dividing line $\kappa = 1/\sqrt{2}$ between type II-superconductors ($\kappa > 1/\sqrt{2}$) and type I-superconductors ($\kappa < 1/\sqrt{2}$)?

Problem 12. Assume $G = B_1$ and $g(\theta) = e^{di\theta}$. We know that $u_{\varepsilon_n} \to u_\star$ having exactly d singularities. Do the singularities of $u_\star$ form a lattice as $d \to +\infty$? Here, we may play with the two parameters: $\varepsilon \to 0$ and $d \to +\infty$. We could first let $d \to +\infty$ (for fixed ε) and then let $\varepsilon \to 0$; alternatively, we could let $\varepsilon \to 0$ and $d \to +\infty$ simultaneously (with some relation between ε and d).

Problem 13. In the framework of Theorem 0.1, let u_ε be a minimizer for E_ε such that $u_{\varepsilon_n} \to u_\star$. Let x_n be a zero of u_{ε_n}, i.e., $u_{\varepsilon_n}(x_n) = 0$. Assume $x_n \to a$ where a is a singularity of $u_\star$. Estimate the rate of convergence $|x_n - a|$ (as $n \to \infty$). Study the blow-up limit of u_{ε_n}, i.e., set

$$v_n(y) = u_{\varepsilon_n}(x_n + \varepsilon_n y)$$

and study the behavior of v_n as $n \to \infty$. This is related to our next problem:

[3] After our work was completed, Problem 9 was answered positively by R. Hardt and F. H. Lin [2]

Problem 14. Study all solutions of the equations

$$-\Delta u = u(1-|u|^2) \quad \text{in } \mathbb{R}^2 \tag{5}$$

having the property that

$$\int_{\mathbb{R}^2} (1-|u|^2)^2 < \infty. \tag{6}$$

Note that for each integer d there exists a solution of (5)–(6) having the form

$$u_d(r,\theta) = e^{di\theta} f_d(r).$$

Are they the **only** solutions of (5)–(6), modulo translation and rotation?

[Some results and further open problems concerning (5)—(6) are presented in H. Brezis, F. Merle and T. Rivière [1]; see also R. M. Hervé and M. Hervé [1] and I. Shafrir [1].]

Problem 15. Replace the energy E_ε and the Dirichlet boundary condition by the appropriate physical expressions arising in type II-superconductors under an applied magnetic field H or in a bucket of superfluid rotated with an angular velocity Ω. Study the asymptotics of minimizers as $\varepsilon \to 0$.

[Some results have been obtained in that direction for superconductors by F. Bethuel and T. Rivière [1].]

Problem 16. Study the minimization problem (2) of the Introduction in the framework of DeGiorgi's Γ-convergence theory.

Problem 17. Assume $G \subset \mathbb{R}^n$, $n \geq 3$, is a smooth bounded domain and fix a (smooth) boundary condition $g : \partial G \to S^{n-1}$ such that $\deg(g, \partial G) \neq 0$. Consider the "energy"

$$E_\varepsilon(u) = \frac{1}{n}\int_G |\nabla u|^n + \frac{1}{4\varepsilon^2}\int_G (|u|^2-1)^2$$

defined on the class of maps $u \in W_g^{1,n}(G;\mathbb{R}^n)$. Study the minimization problem

$$\operatorname*{Min}_{u \in W_g^{1,n}(G;\mathbb{R}^n)} E_\varepsilon(u)$$

and the behavior of its minimizers u_ε as $\varepsilon \to 0$. The main difficulty stems from the fact that $W_g^{1,n}(G;S^{n-1}) = \emptyset$.

Problem 18 (DeGiorgi). Assume v_ε is a solution of the Ginzburg-Landau equation

$$-\Delta v_\varepsilon = \frac{1}{\varepsilon^2} v_\varepsilon (1-|v_\varepsilon|^2) \quad \text{in } G,$$

where G is, for example, a disc. We do **not** fix a boundary condition on ∂G, but we assume instead that, as $\varepsilon \to 0$,

$$E_\varepsilon(v_\varepsilon) \leq K_1 |\log \varepsilon| + K_2.$$

for some constants K_1 and K_2.

Can one conclude that

$$v_{\varepsilon_n} \to v_\star \quad \text{in } C^k_{\text{loc}}(G \setminus \cup_j \{a_j\})$$

for some set $\{a_j\}$ of isolated points? Is $\text{card}(\cup_j \{a_j\})$ controlled by K_1? The answer is not known even for minimizers.

Problem 19. In the framework of Theorem X.1 can one prove that

$$\int_G |\nabla |v_\varepsilon||^2 \leq C \quad \text{as } \varepsilon \to 0?$$

[Recall (see Lemma X.12) that we have only established that

$$\int_G |\nabla |v_\varepsilon||^2 \leq C(|\log \varepsilon| + 1).$$

On the other hand, for minimizers we had the better estimate $\int_G |\nabla |u_\varepsilon||^2 \leq C$; see Theorem IX.4.]

APPENDIX I

Summary of the basic convergence results in the case where $\deg(g, \partial G) = 0$

We recall here, for the convenience of the reader, the main results of F. Bethuel, H. Brezis and F. Hélein [2].

Let $\Omega \subset \mathbb{R}^2$ be a smooth, bounded simply connected domain. Let $g : \partial\Omega \to S^1$ be a smooth map, with

(1) $$\deg(g, \partial\Omega) = 0 .$$

There is a smooth function $\varphi_0 : \partial\Omega \to \mathbb{R}$ such that

$$e^{i\varphi_0} = g \quad \text{on } \partial\Omega.$$

We also denote by φ_0 its harmonic extension in Ω, and we set

(2) $$u_0 = e^{i\varphi_0} \quad \text{in } \Omega.$$

Theorem A.1. *Let u_ε be a minimizer of E_ε in $H^1_g(\Omega;\mathbb{C})$. We have, as $\varepsilon \to 0$,*

(3) $$u_\varepsilon \to u_0 \quad \text{in } C^{1,\alpha}(\overline{\Omega}) \quad \forall\, \alpha < 1,$$

(4) $$\|\Delta u_\varepsilon\|_{L^\infty(\Omega)} \leq C,$$

(5) $$\|u_\varepsilon - u_0\|_{L^\infty(\Omega)} \leq C\varepsilon^2$$

and, for every compact subset $K \subset \Omega$ and every integer k,

(6) $$\|u_\varepsilon - u_0\|_{C^k(K)} \leq C_{K,k}\,\varepsilon^2,$$

$$\left\| \frac{1-|u_\varepsilon|^2}{\varepsilon^2} - |\nabla u_0|^2 \right\|_{C^k(K)} \le C_{K,k}\,\varepsilon^2. \tag{7}$$

Next we consider the case where the boundary condition g also depends on ε. More precisely, we have a family of boundary conditions $g_\varepsilon : \partial\Omega \to \mathbb{C}$ (not necessarily into S^1), and we make the following assumptions

$$\|g_\varepsilon\|_{L^\infty(\partial\Omega)} \le 1, \tag{8}$$

$$\|g_\varepsilon\|_{H^1(\partial\Omega)} \le C, \tag{9}$$

$$\int_{\partial\Omega} (|g_\varepsilon| - 1)^2 \le C\varepsilon^2, \tag{10}$$

$$g_\varepsilon \to g \quad \text{uniformly on } \partial\Omega \tag{11}$$

and

$$\deg(g, \partial\Omega) = 0 \tag{12}$$

(note that, by (10) and (11), g takes its values into S^1).

Theorem A.2. *Let u_ε be a minimizer of E_ε in $H^1_{g_\varepsilon}(\Omega, \mathbb{C})$. Under the assumptions (8)–(12), we have*

$$u_\varepsilon \to u_0 \quad \textit{strongly in } H^1(\Omega), \tag{13}$$

$$u_\varepsilon \to u_0 \quad \textit{uniformly on } \overline{\Omega}, \tag{14}$$

$$u_\varepsilon \to u_0 \quad \textit{in } C^k_{\text{loc}}(\Omega),\ \forall k \tag{15}$$

$$\frac{1-|u_\varepsilon|^2}{\varepsilon^2} \to |\nabla u_0|^2 \quad \textit{in } C^k_{\text{loc}}(\Omega),\ \forall k. \tag{16}$$

The next result is a combination of Theorems A.1 and A.2.

Theorem A.3. *Let u_ε be a minimizer of E_ε in $H^1_{g_\varepsilon}(\Omega, \mathbb{C})$. Assume (8)–(12). Let $x_0 \in \partial\Omega$ and suppose that*

$$g_\varepsilon = g \quad \textit{on } B(x_0, \delta) \cap \partial\Omega,$$

for some $\delta > 0$. Then we also have

$$u_\varepsilon \to u_0 \quad \textit{in } C^{1,\alpha}\left(B(x_0, \tfrac{\delta}{2}) \cap \overline{\Omega} \right). \tag{17}$$

[Strictly speaking, Theorem A.3 has not been proved in F. Bethuel, H. Brezis and F. Hélein [2], but its proof is a direct consequence of the methods developed there.]

Remark A.1. In all the above results we assume that u_ε is a **minimizer** of E_ε; this is used to establish that $u_\varepsilon \to u_0$ strongly in H^1. Suppose now that u_ε is only a solution of the Ginzburg–Landau equation

$$\begin{cases} -\Delta u_\varepsilon = \dfrac{1}{\varepsilon^2} u_\varepsilon (1 - |u_\varepsilon|^2) & \text{in } \Omega, \\ u_\varepsilon = g_\varepsilon & \text{on } \partial\Omega. \end{cases}$$

If, for some reason, we know that u_ε converges strongly to some limit u_0 in H^1, then all the conclusions of Theorems A.1, A.2 and A.3 remain valid. Again, the proofs follow essentially the same arguments as in the proofs of Theorems A.1, A.2 and A.3.

APPENDIX II

Radial solutions

In this appendix we discuss the existence of solutions of the Ginzburg-Landau equation

$$(1) \qquad -\Delta v = \frac{1}{\varepsilon^2} v(1-|v|^2) \quad \text{in } G = B_1$$

satisfying the boundary condition

$$(2) \qquad v = g = e^{di\theta} \quad \text{on } \partial G.$$

The main result is

Theorem A.4. *For every integer $d \geq 1$ there exists a solution of (1)–(2) of the form*

$$(3) \qquad v(r,\theta) = e^{di\theta} f(r)$$

where $f(r) = f_{\varepsilon,d}(r)$ is a function from $[0,1]$ into itself such that $f(0) = 0$.

Proof. Inserting (3) into (1) we are led to

$$(4) \qquad -f'' - \frac{1}{r} f' + \frac{d^2}{r^2} f = \frac{1}{\varepsilon^2} f(1-f^2) \quad \text{on } (0,1).$$

We must solve (4) together with the boundary conditions

$$(5) \qquad f(0) = 0 \quad \text{and } f(1) = 1$$

(the first condition comes from the smoothness of v at 0 while the second condition comes from (2)). The solutions of (4)–(5) arise as critical points of the functional

$$\Phi(f) = \int_0^1 \left[f'^2 r + \frac{d^2}{r} f^2 + \frac{1}{2\varepsilon^2}(f^2-1)^2 r \right] dr.$$

The natural functional space associated to Φ is

$$V = \left\{ f \in H^1_{\text{loc}}(0,1);\ \sqrt{r}\, f' \in L^2(0,1),\ \frac{1}{\sqrt{r}} f \in L^2(0,1) \text{ and } f(1) = 1 \right\}.$$

It is easy to see that

$$V \subset \{f \in C([0,1]);\ f(0) = 0\}$$

and

$$\|f\|^2_{L^\infty} \le \|\frac{1}{\sqrt{r}} f\|^2_{L^2} + \|\sqrt{r} f'\|_{L^2}.$$

Clearly,

$$\operatorname*{Min}_V \Phi \quad \text{is achieved}$$

and this yields a solution of (4)–(5). We may always assume that the minimizer $f \ge 0$ (otherwise replace f by $|f|$). Similarly, we have $f \le 1$; otherwise replace f by $\min\{f, 1\}$ without increasing Φ.

Remark A.2. The uniqueness of a solution of (4)–(5) is proved in R. M. Hervé and M. Hervé [1]. Alternatively one can also use—as was pointed out by I. Shafrir—the method of H. Brezis and L. Oswald [1]. Namely, let f_1 and f_2 be two positive solutions of (4)–(5). Dividing (4) by f and subtracting the corresponding equations we see that

$$-\frac{f_1''}{f_1} + \frac{f_2'}{f_2} - \frac{1}{r}\left(\frac{f_2'}{f_1} - \frac{f_2'}{f_2}\right) = -\left(f_1^2 - f_2^2\right). \tag{6}$$

Multiplying (6) by $r(f_1^2 - f_2^2)$ and integrating over $(0,1)$ yields

$$\int_0^1 \left(f_1' - \frac{f_2}{f_1} f_2'\right)^2 r\,dr + \int_0^1 \left(f_2' - \frac{f_1}{f_2} f_1'\right)^2 r\,dr = -\int_0^1 \left(f_1^2 - f_2^2\right)^2 r\,dr$$

and the uniqueness follows easily.

APPENDIX III

Quantization effects for the equation $-\Delta v = v\,(1 - |v|^2)$ in $\mathbb{R}^2$

We state here, for the convenience of the reader, the main quantization result of H. Brezis, F. Merle and T. Rivière [1].

Theorem A.5. *Assume $v : \mathbb{R}^2 \to \mathbb{C}$ is a smooth function satisfying*

$$(1) \qquad -\Delta v = v(1 - |v|^2) \quad \text{in } \mathbb{R}^2.$$

Then,

$$(2) \qquad \int_{\mathbb{R}^2} (|v|^2 - 1)^2 = 2\pi\, d^2$$

for some integer $d = 0, 1, 2, \ldots, \infty$.

Remark A.3. For every integer d there is a solution v of (1) satisfying (2). In fact, one may find such a v of the form $v(r,\theta) = f(r)e^{di\theta}$. The corresponding equation for f, i.e.,

$$(3) \qquad -f'' - \frac{1}{r}f' + \frac{d^2}{r^2}f = f(1 - f^2) \quad \text{on } (0, \infty)$$

has been studied in detail by R. M. Hervé and M. Hervé [1]. In particular, they show that there is a unique f satisfying $f(0) = 0$ and $f(\infty) = 1$. The uniqueness has also been proved in the Appendix of P. Fife and L. Peletier [1]; still another method consists of using a slight modification of the technique described in Remark A.2.

Remark A.4. If v is a solution of (1) such that $\int_{\mathbb{R}^2}(|v|^2 - 1)^2 < \infty$, one can prove that $|v(x)| \to 1$ as $|x| \to \infty$, in the usual sense. In particular, $d = \deg(v, S_R)$, where S_R is a circle of radius R, is well defined for R sufficiently large. One can show that d satisfies (2).

APPENDIX IV

The energy of maps on perforated domains revisited

In this appendix we present another result of H. Brezis, F. Merle and T. Rivière [1]. It provides a different perspective than in Chapter II to the question of lower bounds for the energy of maps $u : \Omega \to \mathbb{C}$ where Ω is a domain with holes and $|u| \geq a > 0$ in Ω. We have not used it in the book, but it can be helpful in order to derive properties of u_ε, a minimizer of E_ε. We will show, for example, how to prove easily that $\int_G |\nabla u_\varepsilon|^p \leq C_p \ \ \forall p < 2$, as $\varepsilon \to 0$.

Let B_R be the disc of radius R centered at 0. Let $p_1, p_2, \dots, p_m$ be points in B_R such that

$$|p_j| \leq R/2 \qquad \forall j \tag{1}$$

and

$$|p_j - p_k| \geq 4R_0 \qquad \forall j, k,\ j \neq k, \tag{2}$$

so that, in particular, $R_0 \leq R/4$.

Set

$$\Omega = B_R \setminus \cup_{j=1}^m B(p_j, R_0)$$

and let u be a (smooth) map from Ω into $\mathbb{C}$.

We make the following assumptions:

$$0 < a \leq |u| \leq 1 \quad \text{in } \Omega, \tag{3}$$

$$\frac{1}{R_0^2} \int_\Omega (|u|^2 - 1)^2 \leq K \tag{4}$$

for some constants a and K.

Assumption (3) implies that

$$d_j = \deg(u, \partial B(p_j, R_0))$$

is well defined and we consider the "reference map"

$$u_0(z) = \left(\frac{z - p_1}{|z - p_1|}\right)^{d_1} \left(\frac{z - p_2}{|z - p_2|}\right)^{d_2} \cdots \left(\frac{z - p_m}{|z - p_m|}\right)^{d_m}$$

Theorem A.6. *Assume (1)–(4), then*

$$\int_\Omega |\nabla u|^2 \geq \int_\Omega |\nabla u_0|^2 - C\|d\|^2 m^2 \tag{5}$$

where $\|d\| = \sum_j |d_j|$ *and* C *is a constant depending only on* a *and* K.

More precisely, if we set $\rho = |u|$, *then there is a well defined (singlevalued) function* $\psi : \Omega \to \mathbb{R}$ *such that*

$$u = \rho u_0 e^{i\psi} \quad \text{in } \Omega$$

and we have

$$\int_\Omega |\nabla u|^2 \geq \int_\Omega |\nabla \rho|^2 + \int_\Omega |\nabla u_0|^2 + \frac{a^2}{2}\int_\Omega |\nabla \psi|^2 - C\|d\|^2 m^2, \tag{6}$$

where C *depends only on* a *and* K.

For the convenience of the reader we describe the argument following the presentation of H. Brezis, F. Merle and T. Rivière [1].

The proof relies on the following simple:

Lemma A.1. *Given a function* ψ *defined in* $B_{2R_0} \setminus B_{R_0}$, *there is an extension* $\overline{\psi}$ *of* ψ *defined in* B_{2R_0} *such that*

$$\int_{B_{2R_0}} |\nabla \overline{\psi}|^2 \leq C \int_{B_{2R_0} \setminus B_{R_0}} |\nabla \psi|^2 \tag{7}$$

where C *is some universal constant.*

Proof. By scaling we may always assume that $R_0 = 1$ and by adding a constant to ψ we may also assume that

$$\int_{B_2 \setminus B_1} \psi = 0.$$

Poincaré's inequality implies that

$$\int_{B_2 \setminus B_1} |\psi|^2 \leq C \int_{B_2 \setminus B_1} |\nabla \psi|^2.$$

We may then extend ψ inside B_1 by a standard reflection and cut-off technique.

Proof of Theorem A.6. Set $\rho = |u|$. We may write, locally in Ω (but not globally in Ω),

$$u = \rho e^{i\varphi}$$

and then

$$|\nabla u|^2 = |\nabla \rho|^2 + \rho^2 |\nabla \varphi|^2. \tag{8}$$

Similarly, we may write, locally in Ω,

$$u_0 = e^{i\varphi_0}$$

with $|\nabla u_0| = |\nabla \varphi_0|$ and

$$\nabla \varphi_0(z) = \sum_j \frac{d_j V_j(z)}{|z - p_j|} \tag{9}$$

where $V_j(z)$ is the unit vector tangent to the circle of radius $|z - p_j|$, centered at p_j,

$$V_j(z) = \left(-\frac{y - p_j}{|z - p_j|}, \frac{x - p_j}{|z - p_j|}\right). \tag{10}$$

It is convenient to introduce the function ψ **globally** defined on Ω by

$$u = \rho\, u_0\, e^{i\psi}. \tag{11}$$

Thus, we have

$$|\nabla u|^2 = |\nabla \rho|^2 + \rho^2 |\nabla \varphi_0 + \nabla \psi|^2 \tag{12}$$

and consequently

$$\int_\Omega |\nabla u|^2 \geq \int_\Omega |\nabla \rho|^2 + \int_\Omega |\nabla u_0|^2 + \int_\Omega a^2 |\nabla \psi|^2 - X \tag{13}$$

with

$$X = \int_\Omega (1 - \rho^2)|\nabla u_0|^2 + \int_\Omega 2(1 - \rho^2)\nabla \varphi_0 \cdot \nabla \psi - \int_\Omega 2\nabla \varphi_0 \cdot \nabla \psi.$$

We write $X = X_1 + X_2 + X_3$ and we shall estimate each term separately.

Estimate of X_1. We have

$$|\nabla u_0| \leq \sum_j \frac{|d_j|}{|z - p_j|} \leq \|d\| \sum_j \frac{1}{|z - p_j|}, \tag{14}$$

so that

$$\|\nabla u_0\|_4 \leq \|d\| \sum_j \left\| \frac{1}{z - p_j} \right\|_4 \leq \|d\| m \left(\frac{\pi}{R_0^2}\right)^{1/4} \tag{15}$$

Hence, by Cauchy-Schwarz, (15) and (4) we obtain

$$|X_1| \le K^{1/2}\|d\|^2 m^2 \pi^{1/2}. \tag{16}$$

Estimate of X_2. We have, from (9),

$$|\nabla\varphi_0| \le \frac{m\|d\|}{R_0} \tag{17}$$

and thus, by Cauchy-Schwarz and (17), we find

$$|X_2| \le 2\int_\Omega (1-\rho^2)|\nabla\varphi_0|\,|\nabla\psi| \le 2K^{1/2}m\|d\|\,\|\nabla\psi\|_2. \tag{18}$$

Estimate of X_3. We have

$$\int_\Omega \nabla\varphi_0\cdot\nabla\psi = \sum_j d_j \int_\Omega \frac{V_j\cdot\nabla\psi}{|z-p_j|}.$$

We extend ψ inside each disc $B(p_j,R_0)$ using Lemma A.1 and we write, for each j,

$$\int_\Omega \frac{V_j\cdot\nabla\psi}{|z-p_j|} = \int_{B_R\setminus B(p_j,R_0)} \frac{V_j\cdot\nabla\overline{\psi}}{|z-p_j|} - \sum_{k\neq j}\int_{B(p_k,R_0)} \frac{V_j\cdot\nabla\overline{\psi}}{|z-p_j|}. \tag{19}$$

Note that, for $k\neq j$

$$\left|\int_{B(p_k,R_0)} \frac{V_j\cdot\nabla\overline{\psi}}{|z-p_j|}\right| \le \frac{1}{R_0}\int_{B(p_k,R_0)} |\nabla\overline{\psi}| \tag{20}$$

and thus, by Cauchy-Schwarz and Lemma A.1,

$$\left|\sum_{k\neq j}\int_{B(p_k,R_0)} \frac{V_j\cdot\nabla\overline{\psi}}{|z-p_j|}\right| \le C(m-1)\|\nabla\psi\|_2 \tag{21}$$

for some universal constant C.

Finally we observe that

$$\int_{S_r(p_j)} V_j\cdot\nabla\overline{\psi} = \int_{S_r(p_j)} \frac{\partial\overline{\psi}}{\partial\tau} = 0$$

for every $r\in(0,R-|p_j|)$. It follows that, with $\rho_j = R-|p_j|$, we have

$$\left|\int_{B_R\setminus B(p_j,R_0)} \frac{V_j\cdot\nabla\overline{\psi}}{|z-p_j|}\right| = \left|\int_{B_R\setminus B(p_j,\rho_j)} \frac{V_j\cdot\nabla\overline{\psi}}{|z-p_j|}\right| \le \frac{1}{\rho_j}\int_{B_R\setminus B(p_j,\rho_j)} |\nabla\psi|$$
$$\le \frac{1}{\rho_j}\|\nabla\overline{\psi}\|_2(\pi R^2 - \pi\rho_j^2)^{1/2}.$$

Hence we obtain

$$\left|\int_{B_R\setminus B(p_j,R_0)} \frac{V_j \cdot \nabla\overline{\psi}}{|z-p_j|}\right| \le C\|\nabla\psi\|_2. \tag{22}$$

Combining (19), (21), and (22) we are led to

$$|X_3| \le Cm\|d\|\ \|\nabla\psi\|_2. \tag{23}$$

Putting together (16), (18) and (23) we find

$$\begin{aligned}|X| &\le C\,K^{1/2}\|d\|^2m^2 + \|d\|m\,\|\nabla\psi\|_2(2K^{1/2}+C)\\ &\le \frac{1}{2}a^2\|\nabla\psi\|_2^2 + \frac{\|d\|^2m^2}{a^2}(4K+C).\end{aligned}$$

Going back to (13) we obtain

$$\int_\Omega |\nabla u|^2 \ge \int_\Omega |\nabla\rho|^2 + \int_\Omega |\nabla u_0|^2 + \frac{a^2}{2}\int_\Omega |\nabla\psi|^2 - \frac{\|d\|^2m^2}{a^2}(4K+C)$$

where C is some universal constant. This is the desired conclusion (6).

An application. We will show how to prove that if u_ε is a minimizer of E_ε then

$$\int_G |\nabla u_\varepsilon|^p \le C_p \quad \forall p<2, \quad \text{as } \varepsilon\to 0. \tag{24}$$

Of course we already know this fact even for non-minimizing solutions of the Ginzburg-Landau equation. However the proof here is simpler and it yields a better estimate, namely

$$\|\nabla u_\varepsilon\|_{w\text{-}L^2} \le C \quad \text{as } \varepsilon\to 0, \tag{25}$$

where $\|\ \ \|_{w\text{-}L^2}$ denotes the weak-L^2 (i.e., Marcinkiewicz) norm.

We use the notation of Chapter V and we apply Theorem A.6 with $B_R = B(a_j,\eta)$ and $R_0 = \lambda\varepsilon_n$. Here the holes are $B(x_i^{\varepsilon_n},\lambda\varepsilon_n)_{i\in\Lambda_j}$. In Ω_j we write

$$u_{\varepsilon_n} = \rho_{\varepsilon_n} w_{\varepsilon_n} e^{i\psi_{\varepsilon_n}} \tag{26}$$

where w_{ε_n} is the "reference map"

$$w_{\varepsilon_n}(z) = \prod_{i\in\Lambda_j}\left(\frac{z-x_i^{\varepsilon_n}}{|z-x_i^{\varepsilon_n}|}\right)^{d_i}$$

In view of Theorem A.6 we have

$$\int_{\Omega_j} |\nabla u_{\varepsilon_n}|^2 \geq \int_{\Omega_j} |\nabla \rho_{\varepsilon_n}|^2 + \int_{\Omega_j} |\nabla w_{\varepsilon_n}|^2 + \frac{1}{8}\int_{\Omega_j} |\nabla \psi_{\varepsilon_n}|^2 - C \tag{27}$$

and by Corollary II.1 we know that

$$\int_{\Omega_j} |\nabla w_{\varepsilon_n}|^2 \geq 2\pi |\kappa_j|\, |\log(\eta/\varepsilon_n)| - C. \tag{28}$$

Combining (27) and (28) with the upper bound of Theorem III.1 we see that

$$\int_{\Omega_j} |\nabla \rho_{\varepsilon_n}|^2 + \int_{\Omega_j} |\nabla \psi_{\varepsilon_n}|^2 \leq C. \tag{29}$$

From (26) we infer that

$$|\nabla u_{\varepsilon_n}(z)| \leq C\left(|\nabla \rho_{\varepsilon_n}(z)| + |\nabla \psi_{\varepsilon_n}(z)| + \sum_{i\in\Lambda_j} \frac{1}{|z - x_i^{\varepsilon_n}|}\right) \tag{30}$$

Recall that

$$\int_{B(x_i^{\varepsilon_n}, \lambda\varepsilon_n)} |\nabla u_{\varepsilon_n}|^2 \leq C. \tag{31}$$

We easily deduce (24) and (25) from (29), (30) and (31).

Bibliography

A. Abrikosov [1], *On the magnetic properties of superconductors of the second type*, Soviet Phys. JETP **5** (1957), 1174–1182.

L. Almeida [1], *Extremely discontinuous generalized harmonic maps into* S^1, in Proceedings of the first MJI Tohoku University Conference (1993); [2], paper in preparation.

F. Almgren and E. Lieb [1], *Singularities of energy minimizing maps from the ball to the sphere: examples, counterexamples and bounds*, Ann. of Math. **129** (1988), 483–530.

P. Bauman, N. Carlson and D. Phillips [1], *On the zeroes of solutions to Ginzburg-Landau type systems* (to appear).

H. Berestycki and H. Brezis [1], *Sur certains problèmes de frontière libre*, C.R. Acad. Sc. Paris **283** (1976) 1091–1094; [2], *On a free boundary problem arising in plasma physics*, Nonlinear Analysis **4** (1980), 415–436.

F. Bethuel, H. Brezis and F. Hélein [1], *Limite singulière pour la minimisation de fonctionnelles du type Ginzburg-Landau*, C.R. Acad. Sc. Paris **314** (1992), 891–895; [2], *Asymptotics for the minimization of a Ginzburg-Landau functional*, Calculus of Variations and PDE **1**, (1993), 123–148; [3], *Tourbillons de Ginzburg–Landau et energie renormalisée*, C.R. Acad. Sc. Paris **317** (1993), 165–171.

F. Bethuel and T. Rivière [1], *Vortices for a variational problem related to superconductivity* (to appear).

A. Boutet de Monvel-Berthier, V. Georgescu and R. Purice [1], *Sur un problème aux limites de la théorie de Ginzburg-Landau*, C.R. Acad. Sc. Paris **307** (1988), 55–58; [2], *A boundary value problem related to the Ginzburg-Landau model* (to appear).

H. Brezis [1], *Liquid crystals and energy estimates for* S^2*-valued maps*, in Theory and Applications of Liquid Crystals (J. Ericksen and D. Kinderlehrer, eds.), Springer, Berlin and New York, 1987; [2], S^k*-valued maps with singularities*, in Topics in the Calculus of Variations (M. Giaquinta ed.), Lecture Notes in Math., Vol. 1365, Springer, Berlin and New York, 1989, 1–30.

H. Brezis, J. M. Coron and E. Lieb [1], *Harmonic maps with defects*, Comm. Math. Phys. **107** (1986), 649–705.

H. Brezis, F. Merle and T. Rivière [1], *Quantization effects for* $-\Delta u = u(1-|u|^2)$ *in* $\mathbb{R}^2$, C. R. Acad. Sc. Paris **317** (1993), 57–60 and Arch. Rational Mech. Anal. (to appear).

H. Brezis and L. Oswald [1], *Remarks on sublinear elliptic equations*, J. Nonlinear Analysis **10** (1986), 55–64.

H. Brezis and L. Peletier [1], *Asymptotics for elliptic equations involving critical growth*, in Partial Differential Equations and the Calculus of Variations Vol. I (F. Colombini et al., eds.), Birkhäuser, Boston and Basel, 1989.

G. Carbou [1], *Applications harmoniques à valeurs dans un cercle*, C. R. Acad. Sc. Paris **314** (1992), 359–362.

S. Chapman, S. Howison and J. Ockendon [1], *Macroscopic models for superconductivity*, SIAM Review (to appear).

B. Chen and R. Hardt [1], *Singularities for some p-harmonic maps* (in preparation).

P. G. DeGennes [1], Superconductivity of Metals and Alloys, Benjamin, New York and Amsterdam, 1966.

E. DeGiorgi [1], *Sulla convergenza di alcune successioni di integrali del tipo dell'area*, Rendiconti di Matematica **8** (1975), 277–294; [2], *Some remarks on Γ-convergence and least squares method*, in Composite Media and Homogenization Theory (G. DalMaso and G.F. Dell'Antonio, eds.), Birkhäuser, Boston and Basel, 1991.

E. DeGiorgi and T. Franzoni [1], *Su un tipo di convergenza variazionale*, Rend. Mat. Brescia **3** (1979), 63–101.

R. Donnelly [1], Quantized Vortices in Helium II, Cambridge Univ. Press, London and New York, 1991.

Q. Du, M. Gunzburger and J. Peterson [1], *Analysis and approximation of the Ginzburg-Landau model of superconductivity*, SIAM Review **34** (1992), 45–81; [2], *Modeling and analysis of a periodic Ginzburg-Landau model for type-II superconductors*, SIAM J. Appl. Math. (to appear).

C. Elliott, H. Matano and T. Qi [1], *Vector Landau-Ginzburg equation and superconductivity—second order phase transitions* (to appear).

H. Federer [1], Geometric Measure Theory, Springer, Berlin and New York, 1969.

R. Feynman [1], *Application of quantum mechanics to liquid helium*, in Progress in Low Temperature Physics **I**, Chap. 2 (C. Gorter, ed.), North-Holland, Amsterdam, 1955.

P. Fife and L. Peletier [1], *On the location of defects in stationary solutions of the Ginzburg-Landau equation*, Quart. Appl. Math. (to appear).

D. Gilbarg and N. Trudinger [1], Elliptic Partial Differential Equations of Second Order, Grundlehren Math. Wiss. **224**, Springer, Berlin and New York, 1983.

V. Ginzburg and L. Landau [1], *On the theory of superconductivity*, Zh. Èksper. Teoret. Fiz. **20** (1950), 1064–1082; [English translation in Men of Physics: L. D. Landau, I (D. ter Haar, ed.) pp. 138–167, Pergamon, New York and Oxford, 1965].

V. Ginzburg and L. Pitaevskii [1], *On the theory of superfluidity*, Soviet Phys. JETP **7** (1958), 858–861.

M. Grüter [1], *Regularity of weak H-surfaces* J. Reine Angew. Math. **329** (1981), 1–15.

M. Gurtin [1], *On a theory of phase transitions with interfacial energy*, Arch. Rational Mech. Anal. **87** (1985), 187–212.

R. Hardt and F. H. Lin [1], *A remark on* H^1 *mappings*, Manuscripta Math. **56** (1986), 1–10; [2] *Singularities for p-energy minimizing unit vector fields on planar domains* (to appear).

R. M. Hervé and M. Hervé [1], *Etude qualitative des solutions réelles de l'équation différentielle* $r^2f''(r) + rf'(r) - q^2f(r) + r^2f(r)[1 - f^2(r)] = 0, r \geq 0$, *q donné* $\in \mathbb{N}^*$ (to appear).

A. Jaffe and C. Taubes [1], Vortices and Monopoles, Birkhäuser, Boston and Basel, 1980.

D. Kinderlehrer [1], *Recent developments in liquid crystal theory*, in Frontiers in Pure and Applied Mathematics (R. Dautray, ed.), North-Holland, Amsterdam, 1991.

M. Kléman [1], Points, Lignes, Parois, Les Editions de Physique, Orsay, 1977.

R. Kohn and P. Sternberg [1], *Local minimizers and singular perturbations*, Proc. Roy. Soc. Edinburgh **111** (1989), 69–84.

J. Kosterlitz and D. Thouless [1], *Two dimensional physics*, in Progress in Low Temperature Physics, Vol. 7B (D. F. Brewer, ed.), North-Holland, Amsterdam, 1978.

F. H. Lin [1], *Une remarque sur l'application* $x/|x|$, C.R. Acad. Sc. Paris **305** (1987), 529–531.

J. L. Lions and E. Magenes [1], Problèmes aux limites non homogènes, Vol. 1, Dunod (1968); [English translation: Non-homogeneous Boundary Value Problems and Applications, Springer, Berlin and New York, 1972].

L. Modica [1], *The gradient theory of phase transitions and the minimal interface criterion*, Arch. Rational Mech. Anal. **98** (1987), 123–142.

L. Modica and S. Mortola [1], *Un esempio di* Γ^-*-convergenza*, Boll. Un. Mat. Ital. **14** (1977), 285–299.

C. Morrey [1], *The problem of Plateau on a Riemannian manifold*, Ann. of Math. **49** (1948), 807–851; [2], Multiple Integrals in the Calculus of Variations, Grundlehren Math. Wiss. **130**, Springer, Berlin and New York, 1966.

D. Nelson [1], *Defect mediated phase transitions*, in Phase Transitions and Critical Phenomena, Vol. 7 (C. Domb and J. Lebowitz, eds.), Acad. Press, New York and San Diego, 1983.

J. Neu [1], *Vortices in complex scalar fields*, Physica D **43** (1990), 385–406.

P. Nozières and D. Pines [1], The Theory of Quantum Liquids, Vol. II, Addison-Wesley, Reading, MA, 1990.

L. Onsager [1], *Discussion on paper by C. Gorter*, Nuovo Cimento Suppl. **6** (1949), 249–250.

R. Parks (ed.) [1], Superconductivity, Vols 1 and 2, Marcel Dekker, New York, 1969.

L. Pismen and J. Rubinstein [1], *Motion of vortex lines in the Ginzburg-Landau model*, Physica D **47** (1991), 353–360.

O. Rey [1], *Le rôle de la function de Green dans une équation elliptique non linéaire avec l'exposant critique de Sobolev*, C.R. Acad. Sc. Paris **305** (1987), 591–594; [2], *The role of the Green's function in a nonlinear elliptic equation involving the critical Sobolev exponent*, J. Funct. Anal. **89** (1990), 1–52.

J. Rubinstein [1], *Self induced motion of line defects*, Quart. Appl. Math. **49** (1991), 1–10.

J. Sacks and K. Uhlenbeck [1], *The existence of minimal immersions of 2-spheres*, Ann. of Math. **113** (1981), 1–24.

D. Saint-James, G. Sarma and E. J. Thomas [1], Type II Superconductivity, Pergamon Press, New York and Oxford, 1969.

R. Schoen [1], *Analytic aspects of the harmonic map problem*, in Seminar on Nonlinear Partial Differential Equations (S.S. Chern ed.), MSRI Publications **2**, Springer, Berlin and New York, 1984.

R. Schoen and K. Uhlenbeck [1], *A regularity theory for harmonic maps*, J. Diff. Geom. **17** (1982), 307–335; [2], *Boundary regularity and the Dirichlet problem for harmonic maps*, J. Diff. Geom. **18** (1983), 253–268.

L. Schwartz [1], Théorie des Distributions, Hermann, Paris, 1973.

I. Shafrir [1], *Remarks on solutions of* $-\Delta u = u(1-|u|^2)$ *in* $\mathbb{R}^2$, C. R. Acad. Sc. Paris (to appear).

G. Stampacchia [1], Equations Elliptiques du Second Ordre à Coefficients Discontinus, Presses Univ. de Montréal, Montréal, 1966.

P. Sternberg [1], *The effect of a singular pertubation on nonconvex variational problems,* Arch. Rational Mech. Anal. **101** (1988), 209–260.

M. Struwe [1], *Une estimation asymptotique pour le modèle de Ginzburg–Landau,* C. R. Acad. Sc. Paris (to appear); [2] *On the asymptotic behavior of minimizers of the Ginzburg-Landau model in 2 dimensions* (to appear).

R. Temam [1], *A nonlinear eigenvalue problem: The shape at equilibrium of a confined plasma,* Arch. Rational Mech. Anal. **60** (1975), 51–73.

D. Tilley and J. Tilley [1], Superfluidity and Superconductivity, 2d ed., Adam Hilger Ltd., Bristol, 1986.

M. Tinkham [1], Introduction to Superconductivity, McGraw-Hill, New York, 1975.

F. Treves [1], Basic Linear Partial Differential Equations, Acad. Press, New York and San Diego, 1975.

W. Vinen [1], *The detection of a single quantum of circulation in liquid helium II,* Proc. Roy. Soc. A **260** (1961), 218–236.

Y. Yang [1], *Boundary value problems of the Ginzburg-Landau equations,* Proc. Roy. Soc. Edinburgh **114A** (1990), 355–365.

Index

Progress in Nonlinear Differential Equations and Their Applications

Editor

Haim Brezis
Département de Mathématiques
Université P. et M. Curie
4, Place Jussieu
75252 Paris Cedex 05
France
and
Department of Mathematics
Rutgers University
New Brunswick, NJ 08903
U.S.A.

Progress in Nonlinear Differential Equations and Their Applications is a book series that lies at the interface of pure and applied mathematics. Many differential equations are motivated by problems arising in such diversified fields as Mechanics, Physics, Differential Geometry, Engineering, Control Theory, Biology, and Economics. This series is open to both the theoretical and applied aspects, hopefully stimulating a fruitful interaction between the two sides. It will publish monographs, polished notes arising from lectures and seminars, graduate level texts, and proceedings of focused and refereed conferences.

We encourage preparation of manuscripts in some form of TeX for delivery in camera-ready copy, which leads to rapid publication, or in electronic form for interfacing with laser printers or typesetters.

Proposals should be sent directly to the editor or to: Birkhäuser Boston, 675 Massachusetts Avenue, Cambridge, MA 02139

PNLDE1 Partial Differential Equations and the Calculus of Variations, Volume I
Essays in Honor of Ennio De Giorgi
F. Colombini, A. Marino, L. Modica, and S. Spagnolo, editors

PNLDE2 Partial Differential Equations and the Calculus of Variations, Volume II
Essays in Honor of Ennio De Giorgi
F. Colombini, A. Marino, L. Modica, and S. Spagnolo, editors

PNLDE3 Propagation and Interaction of Singularities in Nonlinear Hyperbolic Problems
Michael Beals

PNLDE4 Variational Methods
Henri Berestycki, Jean-Michel Coron, and Ivar Ekeland, editors

PNLDE5 Composite Media and Homogenization Theory
Gianni Dal Maso and Gian Fausto Dell'Antonio, editors

PNLDE6 Infinite Dimensional Morse Theory and Multiple Solution Problems
Kung-ching Chang

PNLDE7 Nonlinear Differential Equations and their Equilibrium States, 3
N.G. Lloyd, W.M. Ni, L.A. Peletier, J. Serrin, editors

PNLDE 8 Introduction to Γ–Convergence
Gianni Dal Maso

PNLDE 9 Differential Inclusions in Nonsmooth Mechanical Problems: Shocks and Dry Frictions
Manuel D. P. Monteiro Marques

PNLDE 10 Periodic Solutions of Singular Lagrangian Systems
Antonio Ambrosetti and Vittorio Coti Zelati

PNLDE 11 Nonlinear Waves and Weak Turbulence with Applications in Oceanography and Condensed Matter Physics
N. Fitzmaurice, D. Gurarie, F. McCaugham, and W. A. Woyczyński, editors

PNLDE 12 Semester on Dynamical Systems: Euler International Math Institute, St. Petersburg, 1991
Kuksin, Lazutkin, and Poeschel, editors

PNLDE 13 Ginzburg-Landau Vortices
F. Bethuel, H. Brezis, and F. Hélein

Printed in the United States
By Bookmasters